AI, Pandemic and Healthcare

Nuoya Chen

Former HEART-ITN Project
University of Macerata
Macerata, Italy

CRC Press
Taylor & Francis Group
Boca Raton London New York

CRC Press is an imprint of the
Taylor & Francis Group, an **informa** business

A SCIENCE PUBLISHERS BOOK

Cover credit: Stability AI generated image and Canva was used to design the cover.

First edition published 2024
by CRC Press
2385 NW Executive Center Drive, Suite 320, Boca Raton FL 33431

and by CRC Press
4 Park Square, Milton Park, Abingdon, Oxon, OX14 4RN

CRC Press is an imprint of Taylor & Francis Group, LLC

Library of Congress Cataloging-in-Publication Data (applied for)

ISBN: 978-1-032-42128-5 (hbk)
ISBN: 978-1-032-42129-2 (pbk)
ISBN: 978-1-003-36132-9 (ebk)

DOI: 10.1201/9781003361329

Typeset in Palatino Linotype
by Radiant Productions

Preface

The world has changed dramatically since I started conducted research in the field of digital healthcare.

The medical community, after the going through the over-promised electronic healthcare records system upgrading, was not enthusiastic regarding the use of digital healthcare solutions and what it may bring before the Covid-19 global pandemic.

Covid-19 has strained the healthcare system and drained the available resources. The waiting time for patients, particularly for those with non-acute and chronic diseases has grown exponentially. This has made the concept of preventive healthcare business model focusing on enabling patients to manage their own health conditions popular and welcoming.

The traditional pay-for-service model does not apply anymore to a healthcare system which needs a constant influx of patients with potential long-term symptoms such as long-Covid.[1] An aging population, with the increasing likelihood of developing chronic diseases (such as diabetes, cardiovascular diseases, cancer, neurological diseases and infectious diseases) can benefit from AI. AI can assist in getting to know the disease better, offering more personalized treatment for patients, and eventual monitoring and control of the disease.

A value-based healthcare model is urgently needed, wherein diagnosis, treatment, and recovery services and solutions are evaluated to create common evaluation standards for medical services and cost coverage. A value-based healthcare model, one where healthcare costs are covered based on the effectiveness of services and medication, is required.

The wide use of AI tools in image/voice recognition and multiple module combinations (for instance, digital twins) allows faster and more efficient diagnosis, particularly for chronic diseases such as Alzheimer's Disease, monitoring and prevention diabetes in pregnant women, nutrition monitoring and apnea. Technology companies are collaborating with governments, hospitals and clinics to develop algorithms aimed

---

[1] According to CDC, "long-COVID" is broadly defined as signs, symptoms, and conditions that continue or develop after acute COVID-19 infection.

at reducing the stress for doctors in their daily work and promoting diagnosis, treatment and recovery efficiency for patients.

This book addresses the challenges towards implementing telehealth solutions in Europe and China. Currently, there is limited use of AI to solve healthcare related problems, due to limitations such as lack of quality data, data interoperability issues and lack of a tangible business model. The author tries to address the challenge for the implementation of telehealth solutions by determining the demand of telehealth solutions in the selected European economies and China (Chapter 1), analyzing the emerging business models for telehealth solution ecosystems in China (Chapter 2), and answering the questions as to how AI will change the scenery for healthcare in the post-Covid 19 world (Chapter 3) and how COVID 19 may change the perspective for each stakeholder when it comes to telehealth solutions (Chapter 4).

Chapter 1 and Chapter 2 form the theoretical background for the empirical work in Chapter 3 and Chapter 4. The book addresses four research questions, namely "Which societal and social-economic unmet needs can Internet of Healthcare Things help to resolve?", "What are the business models innovated for tech companies in China for the smart health industry?", "What are the facilitators and hurdles for implementing telehealth solutions in the post-Covid world?", and "Have stakeholders changed their perspectives regarding the use of telehealth solutions?"

Both qualitative study and quantitative analysis have been performed based on data collected by in depth interviews with stakeholders, and focus group study with potential users (rural and urban residents in Beijing) for telehealth solutions.

The digital platform framework has been used in Chapter 2 as the theoretical framework. Chapter 3 presented a summary of the institutional stakeholder interviews with the interview notes attached whereas Chapter 4 summarizes the focus group study results with individual user interviews.

Telehealth solutions have a great potential to fill in the gap for lack of community healthcare and ensuring health continuity between the home care setting, community healthcare and hospitals. There is a strong demand for such solutions if they can prove the medical value in managing chronic diseases by raising health awareness and lowering health risks by changing the patients' lifestyle. Analyzing how to realize the value of preventive healthcare by proving the health-economic value of digital health solutions (telehealth solutions) is the focus of this research.

Several hurdles still remain in the path to building trust towards telehealth solutions and using AI in healthcare. The next step in this research can also be extended to addressing such challenges by analyzing how the transparency of algorithms can be improved by disclosing the data source and how the algorithms were built. Further research can be done on data interoperability between the EHR systems and telehealth solutions. The medical value of telehealth solutions can improve if doctors can interpret data collected from telehealth solutions; moreover, if doctors are able to make diagnosis and provide treatment, and adjust healthcare management plans based on such data, telehealth solutions can be included in insurance packages, thus making them more accessible.

Contents

Chapter 1

Telehealth Solution Market Demands in China and in Europe (Germany, Denmark and Italy)

1. Introduction

The pandemic has changed the perspective of the healthcare community towards digital healthcare services. Before the pandemic, digital healthcare was seen as an alternative for traditional in-person meetings between doctors and patients. It is difficult to convince insurance companies, governments and users of telehealth solutions to trust digital healthcare solutions. After the pandemic, more funding, investments and attention are flowing into the field of IoT and healthcare, driving rapid developments in the healthcare industry into the Web 4.0 era (Lepore et al. 2022).

Healthcare can be costly and inefficient in different stages—from diagnosis to treatment, or from prevention to care at home. With the population aging in Europe and in China, rising healthcare costs and lack of efficient healthcare solutions are posing challenges for the society. The COVID-19 pandemic has drained the healthcare resources from local healthcare facilities, with healthcare resources concentrated on pandemic control and prevention. The consequences are destroying the pre-pandemic healthcare service structure and have offered a rare chance for digital healthcare service providers. Meanwhile, for patients, the waiting time for healthcare services are longer, costs are higher, suggesting an unsustainable healthcare management system.

Meanwhile, AI has evolved to a degree where much more data is required to generate meaningful insights. Healthcare AI has progressed to a stage where multi-modal analysis has become increasingly important for patient diagnosis and efficient treatment (Nvidia 2022, Nature 2022). It is, however, even more difficult to get data and transfer data across borders as data is being seen as the new oil (The Economist 2017). The intricate balance between getting data, processing data and generating insights, and safeguarding business interests and privacy protection is tricky for startups, multinationals, hospitals and governments.

If there is any lesson to be learned from the pandemic, it is that our society has a strong demand for a robust healthcare sector prepared for a crisis such as COVID-19. With climate change, pandemics such as COVID-19 will happen more and more often. With more funding and resources diverted towards healthcare innovation, one may wonder if AI can realize the promise of lowering healthcare cost, enhancing healthcare efficiency and improving healthcare quality for doctors and nurses (Terry 2016). The volume-based healthcare system has proven to be a failure in the long run. Will AI help alleviate the stress for doctors and patients, and lower the cost for hospitals?

Indeed, technology is changing the landscape for healthcare. In the past, the healthcare industry was stuck in the paper and pen age wherein all health records were stored on paper-based dossiers; diagnosis was made based on experience. Artificial Intelligence (AI) allows automation of repetitive tasks, such as image processing, in the healthcare industry, hence saving time and reducing the workload of healthcare service providers. AI also enables doctors to make diagnosis with the assistance of a vast number of patient records in databases, thus improving the efficiency and accuracy of diagnosis. Wearable devices, chatbots and remote monitoring mechanisms make it possible to raise health awareness among patients, thereby enable them to control their own healthcare conditions easily. However, there are challenges in the existing healthcare model. The volume-based model where patients pay based on the number of services they receive will not work in the post-pandemic world.

Nowadays, it has become even more important that the elderly can live independently and comfortably in their homes. Living at home improves health outcomes and reduces the social expenses for healthcare (like time, energy and resources) (Fattah et al. 2017). When the primary level of healthcare fails to serve as a gate keeper and coordinator for healing, telehealth solutions can potentially fill in the gap by providing solutions with telecommunication and telemedicine for patients with no severe conditions or with chronical diseases or even those just in need of health monitoring. Telehealth solutions provide an alternative model for the aging society with the potential of systematic data analysis, benchmarking best practices and optimizing health outcomes with respect to costs.

For consumers, the internet of things (IoT) stands as an opportunity to make the healthcare system more equitable and efficient, particularly for preventative healthcare and chronic disease management. Smart healthcare solutions such as health coaching apps and wearable devices can potentially alter the traditional ways of treating chronic diseases such as diabetes by promoting patient self-management and homecare, thereby enhancing the quality of life.

Direct-to-consumer telehealth devices offer people the chance to monitor heart rate, sleep quality, weight, blood glucose level, blood pressure, and other key health indicators (Apple 2018, Philips 2018, Huawei 2018, Eversense 2019). Telehealth solutions thus present the possibility of removing barriers for maintaining continuous and sufficient healthcare records of and for patients (Tang et al. 2006). Health coaching apps now provide healthy lifestyle services as well, such as workout coaching, medication sessions, instructions on maintaining oral health, and nutritional advices, thereby promoting health self-management.

Telehealth solutions have promoted healthcare management efficiency by significantly lowering patient waiting time and empowering patients with more information. In China, for instance, telehealth platforms such as Tencent Health, Health and Jing Dong Health offer online consultation sessions with doctors via chat boxes, as well as online patient registration and internet hospital services, including online prescription and sale of medications, medication reminder and other chronic disease management functions (Tencent Doctor Work 2018, Alihealth 2018, Jing Dong Online Healthcare 2019).

For the supply side of healthcare services, telehealth solutions can potentially optimize healthcare outcomes with respect to costs by providing systematic data analysis. Patient data interoperability poses significant challenges for realizing the potential of telehealth solutions. One of the problems lies in patient record and medical image sharing among different levels of healthcare providers and patients. Telehealth networks make it possible for patients to retrieve data from different healthcare institutions and gather them onto a single platform (Dameff et al. 2019). Healthcare data communication standards have also been developed recently, making it possible for patients to connect their mobile devices to healthcare organizations through User Interfaces (UIs). Retrieving health records from patient portals at healthcare institutions, onto smart phones, and sending clinical information (such as glucose level as collected by wearables) to healthcare organizations have been automated. For chronically ill patients, patient-controlled portable records would prove greatly beneficial. Patients with electronic healthcare records compatible with different healthcare information systems can easily receive treatment at different institutions and from different doctors.

Smart healthcare solutions, by equipping patients with data, technology and access to expertise, possess the potential of transforming medical research, medical practices and the role of patients in their own healthcare.

Telehealth now has the capability to democratize healthcare for the first time in history (Standford Medicine 2018). The transformation happens on two levels, namely the distribution of data and the ability to generate and apply insights at scale. It equips patients with data, technology and access to expertise, thereby empowering them to manage their own health. On the institutional level, the transformation brought by telehealth solutions essentially means less focus on routine tasks and more energy for the areas creating the most value and satisfaction (Standford Medicine 2018).

To sum up, the ubiquitous use of mobile telehealth devices, the maturation of health information transfer standards and accessible healthcare-related software have been contributing to personalizing healthcare records. Despite the great potential of connected devices, the capability of telehealth solutions to improve patient outcome, reduce cost and improve healthcare efficiency remains debatable (Dameff et al. 2019). In the stage of adoption of electronic patient health records, systems have shown great enthusiasm. Direct patient-facing technology continuously developed by mature technology companies and well-established research institutions suggest that the digital healthcare landscape may be sufficient to trigger wide adoption of smart health solutions (Dameff et al. 2019). Examples of such solutions include the algorithms detecting early signals of kidney failure, developed by Deep Mind and Google Health (FT 2019); the early detection algorithms for breast cancer, developed by MIT (MIT 2019); and the Neuralink brain implant designed for people with brain and spinal injuries or congenital defects such as Parkinson's disease (The Guardian 2019); all of these solutions have boosted significant scientific and business interests. Meanwhile, the wide adoption of smartphones, the upgraded communication standards, and the availability of software, hardware and sensors also lay the foundation for such technological leaps with AI applications in healthcare (Dameff et al. 2019). It takes time to observe the long-term benefits of AI in healthcare in terms of quality improvement, cost reduction and patient outcomes; yet the enthusiasm of early adopters has suggested the potential benefits of adopting smart health solutions.

Despite heavy investments made by the government as well as the private sector in telehealth solutions in the U.S., Europe and China, embedding IoT in different stages of healthcare is an ongoing process. It is challenging for the medical community to make use of the large amounts of data generated by IoT devices (Deloitte 2018). Most physicians even find it difficult to interpret the data presented by wearables and generate valuable insights (Schnall et al. 2016).

Algorithms may be powerful, yet utilizing AI to identify useful information from redundant data and to improve accuracy in identification and prediction with limited amount of training data in the clinical setting is challenging (Minor 2018). Realizing population health management can be challenging with ethics and legal concerns regarding sharing data over third party app providers (Terry 2016). Consumers may be enthusiastic to try out innovative telehealth solutions in the early stages; but the stakeholders in the healthcare system are yet to take upon the cost for implementation of smart healthcare solutions in the long run. Even further testing is required to determine whether telehealth platforms can be useful, sustainable, scalable, and capable of actually improving health outcomes and lowering costs (Dameff et al. 2019). The research question for this part of the chapter therefore derives from the previous discussions as:

- Which societal and socio-economic unmet needs can the Internet of Healthcare Things help resolve?

In this chapter, the first part is devoted to describing the purpose of research, and the research questions. The second part covers the research methodology by briefly describing the reasons for selecting countries and regions in the EU and in China. The paper compares the healthcare systems in rural and urban areas of China, and the healthcare systems in Germany, Denmark and in Italy. The third part discusses the current trends and emerging technologies in smart healthcare. In the fourth part, a summary of the target for telehealth solutions as observed in Europe and in China is provided. The paper continues to discuss policy initiatives for developing solutions for healthcare services with big data, Artificial Intelligence (AI) and Internet of Things (IoT). The aim is to discern similarities in the emerging trends, challenges and features of the European and the Chinese healthcare systems. The concluding part of the chapter discusses the value added by IoT solutions to different phases in healthcare, following the order of healthy living, prevention, diagnosis, treatment and homecare in different healthcare systems.

2. Research Methodology

Primarily, qualitative study has been used in the paper for parts three, four and five; quantitative analysis has been used to describe the market features and consumer preferences in Europe and in China over healthcare and lifestyle choices. Secondary data has been used in this case for the study. To compare the quality of healthcare in China and Europe, key indicators such as insurance coverage, leading causes of death, and life expectancy in Denmark, Germany, Italy, rural China and urban China have been used.

China is a diverse country with different public health situations in rural and urban areas. Urban residents in Tier-1 cities enjoy better medical resources, including a concentration of high-quality hospitals, doctors and reliable pharmacies (WHO 2015, China Development Research Foundation 2019). As for insurance coverage, most employees in urban enterprises, except for migrant workers, are covered by mandatory Employee Basic Medical Insurance (EBMI). EBMI covers basic medical costs and the coverage ratio varies with cities. In addition to EBMI, the Resident Basic Medical Insurance (RBMI) scheme covers the rest of the residents such as students, the elderly, etc. (State Council 2016).

Compared to urban residents, the rural populations have less reliable medical resources to avail. With the fall of the commune system in the reform era, no effective public health system has emerged as a replacement for the bare foot doctors, in rural areas. There are very few good hospitals (tier-3 hospitals), well-educated doctors and reliable pharmacies providing authentic medication in these areas. Rural residents were only covered by the voluntary New Rural Cooperative Medical System (NRCMS) before 2018, limiting their choices in seeking medical resources. The NRCMS has since merged with URBMI, raising the coverage ratio and coverage range for the type of diseases for rural residents (Pan et al. 2016).

With the ongoing urbanization and population mobilization trends intensifying in China, 57.96 percent of the population now lives in urban areas, availing the medical resources in urban areas (World Bank 2019). 92 percent of all patients in China go to public hospitals because hospitals, particularly the large ones in big cities, are supposed to provide the best care. China has no effective primary care system, making hospitals over-stretched (McKinsey 2010). Therefore, in order to identify the unmet needs of the healthcare system in China, it is essential to differentiate between urban and rural areas.

The European healthcare system is generally more equitable. The general health care system in Europe, despite the differences concerning its extent and quality, presents various competences and values (Leichsenring 2004). It is no news that the health care expenditures are decreasing from Northern to Southern European countries. In terms of social care, different traditions and states of systemic development with a general North-South gap can be observed. Nordic countries started developing specific social services in the 1950s and have developed different types of services and institutions for social care, and professional concepts and approaches. On the other hand, in Southern European countries, most social care services are scarce, with a lack of funding and staff. When it comes to aging and care for the elderly, in terms of social services, Northern and Southern European countries mark a large difference (Leichsenring 2004). Because of the differences in healthcare quality and services observed in the past literature, we chose to focus on three significant national cases: Denmark, Germany and Italy.

The reason for choosing Denmark is that it leads in Europe—in terms of digitalization of the healthcare system (Kierkegaard 2013). Denmark has established a national electronic m-health (eHealth) portal which provides access to personal health data from hospitals, general practitioners' offices and municipalities (Ministry of Health Denmark 2017). The Danish eHealth portal aims at providing a more coherent patient experience and facilitating treatment locally, regionally and nationally (WHO 2018).

Germany has been chosen as an example because it represents a Northern European country which has important an economic as well as political role to play in the European Union. Germany is on its way towards the digitalization of its healthcare system (Lovell 2019). Berlin has also become one of the most exciting start-up incubating cities in Europe; therefore, Germany is an interesting example to study for the potential implementation of smart healthcare solutions.

Italy is an example of a Southern European country, the healthcare system of which, owing to its efficiency, currently holds the 4th place in the world (Bloomberg 2018c).

3. Current Trends and Emerging Technologies in Smart Healthcare

3.1 Wearables, monitoring of patients and preventive healthcare

Traditionally, healthcare is a closed-cycle process where information flows unidirectionally—from physicians to patients; research universities serve as the center of innovation and progress in healthcare (Stanford Medicine 2018). Due to the digitalization of patient healthcare records, data flow is now becoming increasingly free. Patients are engaged in more complex data sharing with connected devices and wearables. This gives patients more power to be engaged in their healthcare process.

Wearables today offer patients the opportunity to monitor key parameters such as heart rate, enabling early detection and self-diagnosis of cardiovascular diseases. For instance, Apple Watch, Fitbit, Xiaomi Watch and other smart watches and bands have embedded photoplethysmography (PPG) sensors to monitor the heart rate via transmission and absorption of light against the skin (Allen 2007).

Table 1.1 lists the world's top 5 wearable brands and their health monitoring and coaching functions. Most variables today have basic heart rate monitoring, sleep monitoring and work-out tracking or coaching, and have integrated VO2 monitors and virtual assistants. Only the Apple Watch has an integrated fall detection, ECG diagram generator, nutrition tracking, stress management function, etc. Sadly, all of the World's top five wearable brands belong to American or Chinese companies. Apple and Fitbit are tech companies based in San Francisco, Garmin originated

Table 1.1: World's Top Five Wearable Brands (by shipment) and their health/well-being-related functions; Source: Apple 2019, Xiaomi 2019, Fitbit 2019, Huawei 2018, Garmin 2019.

Manufacturer/Function	Apple	Xiaomi	Fitbit	Huawei	Garmin
Fall Detection	X				
ECG Diagram Generation	X				
Heart Rate Monitor	X	X	X	X	X
Emergency SOS Call	X				
Meditation Coaching/Stress Management	X		X		X
Period Tracking	X		X		
Glucose Tracking	X				
Nutrition Tracking	X				
Water Intake Tracking	X				
Workout Tracking and Coaching	X	X	X	X	X
Sleep Monitoring	X	X	X	X	X
Call & Text	X	X (Reminders)	X (Reminders)	X	X
GPS Tracking	X	X	X	X	X
VO2 Max Tracking	X	X	X	X	X
Wireless Music Play	X	X	X	X	X
Virtual Assistant	X	X	X	X	

in Texas, while Huawei and Xiaomi are based in mainland China; no European brands appear in the list.

Monitoring diets, activities, medication intake, sleep quality and heart rates is important for doctors for precise diagnosis and speedy recovery of patients. Image recognition algorithms can make it faster for doctors to process test images, therefore reducing mistakes. Patients with chronical conditions may find it much easier to adjust their treatments, communicate with healthcare professionals, and perform self-care. For healthy living, smart health solutions already provide a wide range of solutions for controlling stress levels, monitoring and enhancing sleep quality, workout coaching, pregnancy-related functions such as portable ultrasound, and dietary logging.

Smart healthcare solutions have the potential to improve healthcare quality by promoting healthcare efficiency. For hospitals, deploying IoT solutions can help doctors in reducing repetitive workloads and management for better cost control. Automating repetitive tasks such as data entry to the Electronic Health Records Systems (EHRs) and managing medical resources and in/out-flow of patients can greatly reduce costs for hospitals (Stanford Medicine 2018). Smart health solutions can also deliver faster and more precise diagnosis, and reduce burn out risks for physicians with AI assistants providing advice regarding image reading, diagnosis and prescription.

Artificial intelligence has shown great promise in clinical areas such as diabetes, heart disease, cancer, neurological disease, etc. (Rahman et al. 2021). AI has been projected to reduce the cost for healthcare in the United States by $150 billion by 2026 (Accenture 2018, Stanford Medicine 2018). The most valuable new developments herein have been predicted to be robot-assisted surgeries, with a projected economic value of $40 billion. The next most valuable development is virtual nursing assistance, with a projected economic value of $20 billion, and the third most valuable development is administrative work-flow assistance, with a projected economic value of $18 billion (Accenture 2018, Stanford Medicine 2018).

Telehealth solutions provide the prospect of establishing a systematic data processing system to cover each stage of healthcare to perform preventive healthcare tasks. Telehealth devices can perform real-time and synchronized monitoring of biometric and environmental indicators, including temperature, humidity, luminosity and acoustic noises. With the systematic monitoring of human body and its interaction with environment, it is possible to analyze the correlations between the environment and the bio-signals (Rinbeat 2018). Systematic monitoring across a large population can help scientists to recognize significant trends and patterns related to human health and develop algorithms. These algorithms can potentially predict the users' stress and well-being, emotions, cardiopathies, discomfort level, quality of sleep, sports performance and recovery time, attention capacities and concentration level, etc. (Rinbeat 2018).

Ultimately, wide applications of AI and IoT in health, fitness, military, automotive industry, transportation and insurance may emerge. For instance, for military purposes, fitness trackers combined with environmental sensors can help soldiers in predicting attention capacities and adjusting stress levels, hence adapting to war situations. In the field of sports, the system will be able to enhance sports performances, calculate recovery time and evaluate the level of discomfort for users. When it comes to healthcare, the system has the capacity to perform real-time monitoring of the heart and the central autonomic system (Rinbeat 2018).

With healthcare costs on the rise, Internet of Healthcare Things can help in curbing this trend by preventing diseases and promoting a healthy lifestyle. Expenditure on preventive healthcare has long been but a small part of the entire healthcare system; most of the spending is focused on acute and extreme conditions. Internet of Healthcare Things can potentially detect early symptoms of complex conditions such as breast cancer (MIT 2019, Ehteshami Bejnordi et al. 2017). Wearables, by monitoring vital signs, can warn patients of important changes in their biometric signals, thereby urging them to utilize early intervention methods. Population health management can predict the breakout of epidemics, therefore making control efficient. The China National Disease Control Center has already implemented real-time monitoring systems for communicable diseases such AIDS and hepatitis A/B for early intervention.

The use of telehealth solutions is vital for establishing a value-based healthcare system, by establishing system wide monitoring of treatment, outcomes and costs. Using systematic data processing and benchmarking of best practices, telehealth provides a potential solution for combating the rising healthcare cost and lack of access to basic healthcare services.

3.2   New business models for internet of healthcare things developed in China

It is argued that the Internet of Things is more about business play rather than technology (Subramania 2019)—the way business organized for Internet of Things should start from identifying the market needs, and how technology can feed the needs better than the current methodology (Subramania 2019). The more important thing here is to align innovation with market needs and organize products and services with IoT and extend the market into new areas (Subramania 2019).

Currently in China, big tech companies are marching into the healthcare market, trying to offer products and services targeting at the full cycle of care. Tables 1.2 and 1.3 summarize their services and business models. These companies can be categorized into software and hardware companies. The leading players in software are Alihealth, Wedoctor, Jing Dong (JD) Health, Ping An Good Doctor and Xiaohe Medical (backed by Bytedance), whereas the leading hardware producers are Tuya, Xiaomi and Huawei. Among the successful companies, the business models vary. Some of these companies have chosen to establish their own apps and networks, like Alihealth and Ping An. Tencent, however, has invested in its own app as well as in smart health startups like We Doctor Platform, Ding Xiangyuan, etc. Compared to traditional healthcare providers, emerging players have certain advantages. Firstly, the emerging players do not have the legacy of developing and manufacturing medical devices; hence, the baggage free strategy enables them to be more flexible in developing telehealth solutions focusing on a specific disease or disease groups. For instance, the Google team focuses on developing algorithms to help identify hand gestures in short videos uploaded by users to develop algorithms used by robotic hands.

One of the challenges for telehealth solutions lies in the interoperability of electronic health record (EHR) systems. Till today, doctors in Europe typically put medical images on CD-ROM and give the CDs to patients

Table 1.2: Leading Smart Health Service and Product Providers, filtered by the number of users. Source: Technode 2020, Financial Times 2016, Alihealth 2018, Tencent Doctorswork 2018, Bloomberg 2022.

Companies	Wearables	App	Mobile Payment for hospitals	Medical Service and Medicine O2O Sale	Self-Cloud Computing Platform for Developers	Online Medical Education/ Medical Knowledge Verification	AI Appliances in Diagnosis	Smart Home Appliances-Hardware	Internet Hospital/ Investment related to Internet Hospitals	M&A related to/ Association with Insurance Company	Data Interoperability standards consultation
We Doctors Limited Holdings		X	X	X		X	X		X	X	
Tencent		X	X		X	X	X		X	X	
Ping' an Healthcare and Technology Ltd		X	X	X	X		X		X	X	X
Xiaohe Health- Subsidiary of Bytedance		X		X		X			X		
JD Health		X		X	X	X			X	X	
Alihealth-Alibaba		X	X	X	X	X	X		X	X	X
Microsoft		X			X		X				X
Samsung	X	X	X		X		X	X			X
Amazone Healthcare	X	X	X	X	X		X	X	X	X	X
Apple	X	X	X		X		X	X			X
Google	X	X	X	X	X		X	X			X

Table 1.3: Business Models and Profitability of leading connected health product and service providers, filtered by the number of users. Source: Financial Times 2016, Alihealth 2018, Tencent Doctorswork 2018, Bloomberg 2018a, B2B: Business to Business, Q1: Quarter 1, O2O: Online to Offline.

Hardware	Wearable	App	Hospital used Medical Equipment	Cloud Computing Platform for Developers	AI Appliances in Diagnosis	Smart Home Appliances-Hardware	Connected Hospitals	Data Interoperability standards consultation
Huawei	X	X		X		X	X	
Xiaomi	X	X				X		
Simens			X		X		X	X
Philips			X		X		X	X
GE			X		X		X	X
Mindray			X		X		X	

when most laptops no longer carry a CD reader. Hospitals still use fax to share patient records (Stanford 2018). In China, patients usually pay US $0.5 for a blue paper book where the doctors record notes. Patients then get the X-Ray or MRI images printed out and taken away; most patient records are stored in separate electronic health record systems at different hospitals. EHR records are static and usually stay within the institutions that collect them; EHR data stored in different systems creates the interoperability problem. Hospitals generally have no exact standards on the type of data input in the EHR system, with each individual physician and department entering data according to their own preferences. This brings disparity and differences in the quality for EHR stored (Stanford Medicine 2018). For some reason, hospitals see the data of patients as their own assets, often refusing requests from patients to transfer data between organizations.

For patients with chronic diseases, it is difficult to consult with different levels of healthcare service providers with scattered medical records at different medical institutions. There is a great interest among patients to access information from hospital records and carry convenient, patient-controlled portable records. Personal health records are distinct from electronic patient records maintained by the health care system patient portals—personal health records are deposits of clinical data managed by patients (Bates and Wells 2012). These records may contain the same types of data as in hospital maintained medical records (such as medical history, diagnostic test results and clinician documentation) (Dameff et al. 2019). Telehealth solutions offer a change to patients for such services.

Doctors, on the other hand, find EHR input a large part of their daily routine. A study conducted by Stanford Medical School confirmed that EHR inputs contributed to burn out in physicians and has stood in the way of doctor-patient communication (Stanford Medicine 2018). 44% of all physicians surveyed said that the primary purpose of EHR is data storage, with 8% of physicians citing medical factors related to EHR (Stanford Medicine 2018). One third of all the physicians surveyed expressed hopes for financial information to be integrated in the system so that patients would be able to weigh their cost options; the primary concern is still the interoperability problem, where data can become available for all parts of healthcare systems.

Health Level Seven International (HL7.org)[1] has developed Fast Healthcare Interoperability Resources (FHIR), offering a standard data

---

[1] Health Level Seven International (HL7) is a not-for-profit, ANSI-accredited standards developing organization dedicated to providing a comprehensive framework and related standards for the exchange, integration, sharing, and retrieval of electronic health information that supports clinical practice and the management, delivery and evaluation of health services, available at: http://www.hl7.org/about/index.cfm?ref=nav, last accessed on September 2nd 2019.

formatting and an API for exchanging EHR (Stanford Medicine 2018). The FHIR has become popular with the healthcare community. Apple, for instance, has taken use of it and made it possible for consumers with IOS 11.3 beta to have medical information from various organizations organized into one view (Apple 2018). These medical information covers allergies, conditions, immunizations, lab results, medications, procedures and vitals; consumers also receive notifications when data gets updated. By January 2019, 163 hospitals and clinics in the US have connected their EHR portal with Apple (Apple 2019).

In China, the Ministry of Public Health has drafted the 'Health Profile of Basic Architecture and Data Standards (for trial implementation)' (BLED 2010 Proceedings et al. 2010). The new regulation set standards for five types of EHRs—(a) basic personal health information record, (b) disease control record, (c) maternal and child health record, (d) medical services and (e) community health record (BLED 2010 Proceedings et al. 2010). However, there exists no standard 'Community Health Service Information Management System'. Most hospitals build their EHR systems based on the hospital information system (HIS). By 2011, in China, 120 EHR systems were fully in place and being used daily; also, 40 more systems were on their way to completion, and 100 more were in the planning stage among approximately 20,000 hospitals. As of 2014, half of all tertiary hospitals in China use their own EHR system, as do 30% of urban health centers and 20% of rural hospitals (Pacific Ventures 2016).

In Europe, the heterogeneity of the EHR system also became one of the largest barriers for accessing healthcare data across borders (European Commission 2019b). The European Commission started public consultation in 2017 to address the issue. Following the consultation, the EC adopted a recommendation on the exchange of electronic health record formats. Several projects and initiatives have been carried out since then to promote cross border data exchange. The e-Health digital infrastructure framework has been implemented (European Commission 2019b) to facilitate the exchange of two types of data: patient summaries and electronic prescriptions. In the beginning of 2019, Finland and Estonia managed the exchange of patient information. By 2021, 22 member states were expecting to exchange these types of information (European Commission 2019a). In general, the northern European countries were the first to launch digital health initiatives, while the southern European economies only recently started to launch the digital health platform. The market acceptance of smart health solutions, both with institutional customers (government and private sector) and with individual users varies significantly.

In Denmark, 98 percent of all healthcare records are exchanged electronically (Ministry of Health Denmark 2017). Patients can access eSundhed.dk for information about the quality of healthcare services and make an informed choice about hospitals. The other website, Sundhed.dk, provides patients with a few personal services and data such as patient records from hospitals (e-journals), as well as general information on health, diseases and patient rights. With mature IT support, it is relatively easier for doctors to perform tele-medicine treatment to manage chronically sick patients, offer medication adjustments, and advice for diet and exercises. For instance, the recent program for chronic obstructive pulmonary disease (COPD) patients, run in Denmark, has successfully reduced hospital admission rate, with tele-medicine proving to be less time consuming than regular visits to hospitals. The other programs intend to assess the wounds for 70 percent of all patients in Denmark through tele-medicine. Trained nurses tend to the wounds of patients at home or at local health clinics, take a picture and upload the picture to the online patient records for doctors at the hospitals to assess. The program reduces hospital admission time, saves the time for healthcare professionals and transportation time for patients. From the Danish experiences and business model analysis of leading telehealth product and service suppliers, the following can be summarized about developing a successful telehealth system (Ministry of Health Denmark, 2017).

Germany is the biggest healthcare market in Europe (German Trade and Invest 2017). There is a wide acceptance of fitness trackers for lifestyle improvement, such as nutrition monitoring, calorie control and sleep tracking, with 31 percent of the entire German population using wearables to track their biometric signals (German Trade and Invest 2017). The integration of smart health solutions in the primary level of healthcare brings huge market potential. The main hurdle for implementing such solutions in Germany is the strict market regulation (German Trade and Invest 2017). The insurance industry is cooperating with startups and has managed to reimburse the use of several healthcare apps. For instance, one of the largest health insurance firms in Germany, Barmer, has co-developed the app 'Hearing Test' with Mimi Hearing Technology to help youngsters test their hearing ability at home, using headsets. The German Federal Institute for Drugs and Medical Devices (BfArM) oversees the differentiation between medical apps and fitness and life style ones (BfArM 2019). The German Medical Device Act and the European guidelines apply to the classification of the app (German Trade and Invest 2017). Once considered as a medical app, the software and the app are subject to the same rule as regular medical devices. The medical app will then need to follow CE marked and EU guidelines (German Trade and Invest 2017). The German smart health market had a market value of US $2.967 billion in 2017 (German Trade and Invest 2017).

In Italy, there have been attempts to launch the national digital heath strategy since 2014. In 2017, the government launched the 'three-year

Figure 1.1: Smart Health Ecosystem. Source: Kyng et al. 2019.

plan for digitalization of the public administration' (Agency for Digital Italy 2018). This policy initiative aimed at improving healthcare service quality, taking control of waste, improving healthcare cost efficiency and lowering the health service inequality among different regions in Italy (Agency for Digital Italy 2018). There are three projects associated with the digital health initiative proposed by the 'three-year plan'. The plan aims to build an EHR system which serves as the infrastructure for digital health solutions, a central e-health hub which serves as the sole interaction platform between patients and the public health administration, and a telemedicine system (Agency for Digital Italy 2018). The projects also aim to facilitate cross border data transfer of patient summaries and the e-prescriptions, following the EC recommendation (Agency for Digital Italy 2018).

In China, for instance, regional hospital networks embedded with larger IoT networks such as smart city networks, centralized data processing centers and cloud platforms now allow doctors to check patient electronic healthcare records online. Internet hospitals also allow patients to start online consultation for second time visits (Wuzhen Internet-based hospital 2019). Cloud computing has also made it possible to perform pandemic disease management, predict an epidemic outburst and analyze the contributing factors for chronical diseases (China News 2018). Ping An Health Cloud Computing platform has also expanded from collecting and organizing clinical data to providing medical resources management and medical cost management services by incorporating data from private and public medical insurance systems (China News 2018).

The advancement of technology in healthcare has been reducing costs for healthcare, providing easier access to healthcare, personalizing healthcare services, boosting more precise and immediate treatment, and playing a greater role in doctor-patient relationship in general (Stanford Medicine 2018). One of the positive changes brought by smart healthcare is the reduced patient waiting time. Patient waiting time has been identified as one of the key indicators by the WHO for a responsive healthcare system (WHO 2003).

For instance, in China, the waiting time mainly occurs at the hospital reception and the admissions window, and between the appointment time and the time at which patients are attended by doctors (Sun et al. 2017). Usually, Chinese patients need to endure long waiting times in hospitals for registration, check-ups and payment; the observable change with telehealth solutions is a significant reduction in waiting time for treatment. The mobile payment and patient registration system connected to Wechat, a popular Chinese social medial software available for smartphones and computers, with more than 1 billion users (Forbes 2018), has significantly lowered the waiting time for patients and promoted accessible healthcare (Forbes 2014).

The mobile payment system has integrated patient registration, payment, and prescription distribution services. With Wechat, more than 13,300 hospitals now allow patients to pay with their mobile devices; more than 22,800 hospitals offer an online appointment system, and more than 38,000 hospitals publish healthcare related information via Wechat public accounts (Shen 2020, McKinsey 2019).

Wearables with heart rate monitoring functions make it possible for patients with potential cardiovascular diseases to get prompt treatment. Recently, a patient with intermittent palpitations in his 70s observed paroxysms of tachycardia at rest on his fitness tracker. The patient then purchased an AliveCor Kardia device approved by the US Food and Drug Administration (FDA) to get single-channel electrocardiographic images on his smart phone. He recorded the irregular heartbeats with retrograde P waves. The patient's self-diagnosis was confirmed by subsequent 12 lead electrocardiographic recordings and he received an appropriate treatment (lp, JAMA 2019). The newest model of Apple Watch is able to generate electrocardiographic recordings with the back and the bezel of the watch serving as lead I electrocardiographic bi-poles (lp, JAMA 2019). Apple Watch 4 is designed to detect occult atrial fibrillation (AF) (Apple 2018). Wearables now have the potential to improve the diagnosis of cardiovascular diseases such as sporadic or occult AF. However, it remains challenging for physicians to analyze and interpret the information provided by the smart devices of patients. The reason is that the devices may generate false alarms and result in unnecessary further testing for the patients; one recent study involving 100 patients with cardiovascular AF conditions showed that the device's algorithm categorized 34 percent of all recordings as unclassified even under direct observation (Bumgarner et al. 2018b). Applying diagnosis based on recordings generated by smart devices poses the risks of misinterpretation and inappropriate results, which can be problematic if devices are used in a population with low prevalence of diseases.

To summarize, telehealth solutions bear a huge market value and a great potential to improve efficiency and save costs in each stage of healthcare. There are still challenges towards implementing smart healthcare solutions, the biggest one being the lack of interoperability. Further challenges towards the large-scale implementation of telehealth solutions in the healthcare process lie in data management, cybersecurity, regulation and financing.

4. Unmet needs of Healthcare Systems in China and Europe

4.1 *The general European healthcare system: overview and implications for telehealth*

The European healthcare system has been identified to be consisting of three types of healthcare system clusters:

Type I: a health service provision-oriented system which provides a large number of healthcare service providers and offers patients free access to doctors;

Type II: a universal coverage system whereby access to healthcare is considered as a citizenship right. Therefore, equal access for healthcare services is recognized as a social citizenship right. This puts equal access to healthcare services at higher priority than free access and freedom of choice for patients.

Type III: a low-budget restricted access system where patients have limited access to the healthcare system due to a high percentage of out-of-pocket payments and the fact that patients have to sign up for a family doctor and endure long waiting times (Wendt 2009).

Among the three countries considered in our case study, Germany belongs to the first cluster, where the healthcare system is characterized by a high level of health expenditures and the system receives a substantial amount of public funding (OECD 2017); the portion of out-of-pocket patient payment is moderate. There are also high levels of outpatient care and a moderate level of inpatient care. Furthermore, many autonomous self-employed doctors are present, offering a high level of choice to patients.

Denmark and Italy belong to the universal coverage group, wherein the healthcare system incurs a medium level of total health expenditure. The share of the public health spending is high, while the out-of-pocket payment ratio is moderate. Compared to the previous group, the number of inpatient healthcare providers are similar, but the number of outpatient provider is low; access to healthcare services is highly regulated, while doctors themselves face a high level of regulation for additional income opportunities (Wendt 2009).

Figure 1.3 describes the healthcare spending in Denmark, German and Italy from 2012 to 2016. This correlates with the population aging trends in the EU as shown in Figure 1.2, Figure 1.3 and Figure 1.4. The average life expectancy at birth in the European Union was 80.62 years in 2016, with age dependency ratio reaching 54.39% in 2017 (World Bank 2018). Figure 1.2 shows the age dependency ratios in the EU and China. Both economies are marching into the stage of aging, with aging following a more intensive pace in the former.

Strong primary care is supposed improve the capacity of a country to establish a responsive, high-quality and cost-effective healthcare system. The primary level of healthcare structure consists of primary care governance, economic conditions for primary care, and primary care workers development (Kringos et al. 2013).

The primary care process is concerned with primary healthcare accessibility, coverage scope for primary level of healthcare, continuity of primary level and coordination of primary care (Kringos et al. 2013). In the EU 27 member countries, Belgium, Denmark, Estonia, Finland, Lithuania, the Netherlands, Portugal, Slovenia, Spain and the UK are identified to be equipped with a strong primary care system (Kringos et al. 2013). Germany shows a weak coordination role of primary level healthcare, while Italy shows weak coverage scope in the primary level of healthcare. GPs in western European countries generally earn more than those in the eastern European countries.

Countries with strong primary care usually exhibit the following characteristics:

First of all, the primary level of healthcare is the main point of entry for the rest of the healthcare system. Most of the countries have enacted national policies to ensure access to primary care, both geographically and financially. The first point of contact to the healthcare system can also be nurses, where a supplementary policy can enhance the accessibility of the healthcare system.

Secondly, the primary care system assumes a medical advocacy role for individual patients, monitoring prevention, diagnosis, treatment and other follow up activities.

Thirdly, the primary level of healthcare takes up the coordination role within and outside the primary level of care.

The fourth characteristic of countries with a strong healthcare system is a strong commitment towards universal access to primary level healthcare; all countries have lowered the co-payment ratio for primary healthcare as much as possible; policies ensuring accessibility for the lowest income group, the elderly, patients with chronical conditions or disabilities,

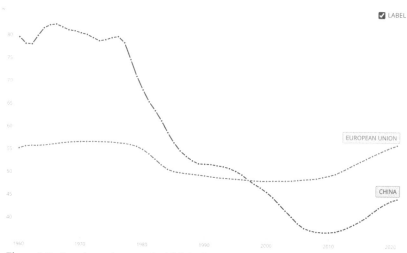

Figure 1.2: Age dependency ratio (old) in the European Union and China, 1960–2021. Source: World Bank 2022.

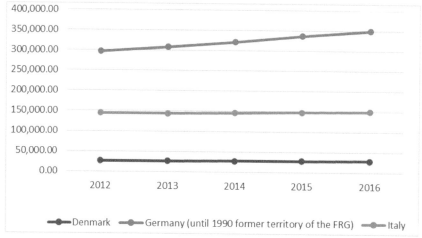

Figure 1.3: Healthcare expenditures per capita in Germany (top), Italy (middle) and Denmark (bottom), 2012–2016. Source: Eurostat 2021.

children and pregnant women have been designed. For access to medication, a co-payment system has been developed with the insurance system offering deductibles.

The European healthcare policy making is heavily guided by the principle of subsidy (Jakubowski and Busse 2002). With the global financial crisis in 2008, the governments in Europe have been struggling between austere measures which cut government budgets and increased healthcare costs due to the increasing segment of the aging population with chronical diseases. With the tumbling healthcare budget and growing healthcare

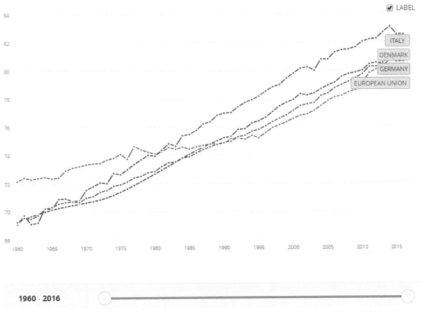

Figure 1.4: Life Expectancy in Germany, Denmark and Italy, along with the EU average.
Source: World Bank 2022.

demand from the population, policy makers in Europe are trying to find a way to improve healthcare quality while taking control of healthcare budgets. Several European countries are trying to implement the value-based healthcare system, in an attempt to look at the healthcare system as a whole. However, there are several challenges in the implementation of the value-based healthcare model: (a) lack of detailed data on healthcare outcomes across the care cycle, (b) growing pressure for new pricing models, especially in terms of medications and medical devices, (c) need for mechanisms to ensure that the most vulnerable group of the society also has access to healthcare (Economist Intelligence Unit 2016). So far in Europe, there have been many pilot programs running at individual healthcare institutions for value-based healthcare models; there has also been greater focus on their effects on the spending on pharmaceutical and technology by the general population. There is even greater collaboration for developing the European Network for Assessing Health Technology Assessment for better evaluation of the cost effectiveness, for policy makers to make precise decisions (Economist Intelligence Unit 2016). There is also a need to improve government-industry collaboration to maximize the value of investments in health technologies. Furthermore, there is a need to ensure accessibility of the healthcare system, a problem for countries with weak primary level healthcare.

The Internet of Healthcare Things can potentially monitor patients closely over the whole care cycle, and therefore offers a good opportunity to identify best practices in the healthcare system to establish benchmarking of healthcare services, make sure that all team members follow the same protocol, share data across different levels of healthcare so as to ensure the continuity of healthcare services, and promote access to healthcare services by providing online consultation. By collecting large amounts of data and using algorithms to detect patterns in the data, it is easier to detect early stages of chronical diseases before the symptoms even appear (Kaminsky 2019); it is also easier to provide better treatment with the established protocol and follow up on patients as to whether they follow the clinical pathway. Therefore, integrating telehealth solutions with the value-based healthcare system can potentially lower healthcare cost and improve healthcare quality and efficiency.

4.1.1 Demark

The Danish healthcare system has traditionally led the European digital healthcare scene in the level of digitalization, the efficiency of the healthcare system, and healthcare cost reimbursement. In such an open market, the Danish healthcare system is one which welcomes external technology. In 2018, the Danish government carried out a 4-year plan to implement better digital healthcare strategy. The plan projects that the population over 75 years old is growing and will double in the next 30 years—from 9.8% to 14.4%. The plan analyzes the tendency of the elderly to use internet—by 2016, 86% of all elderly users have become internet users.

The Danish healthcare system operates with three administrative levels—national, provincial, and local. The state is in charge of regulatory and supervisory functions; the five regional administrative care systems are responsible for hospitals, GPs and psychic care. The municipalities take care of a number of general practitioner services and elderly care (Ministry of Health Danmark 2017). Normally, all healthcare services are paid by general taxes and are supported by the central government's grant, reimbursement and equalization schemes. Public finance provides 84 percent of all healthcare expenditures, while the rest 16 percent was funded through patient co-payment schemes in 2015. Healthcare costs accounted for 30 percent of the total public expenditures and 10.6 percent of GDP in 2015, which is higher than the average 9 percent of GDP in OECD countries (Ministry of Health Danmark 2017).

All residents in Denmark have access to healthcare, with most of the services offered for free. National legislation ensures that diagnosis and treatment are offered within certain time limits and patients have the right to a free choice for hospitals. Residents in Denmark may also seek treatment from abroad if treatment has not been offered in Denmark, upon approval;

treatments received in other European Union (EU)/European Economic Area (EEA) countries are also entitled for reimbursement. Patients' complaints and compensation for injuries incurred in the treatment procedures are guided by a specific set of rules (Ministry of Health Danmark 2017). Every year, over 25,000 patients are invited to participate in patient empowerment programs through which the differences in patient experiences are compared, inputs for quality improvement are collected and development for patient experiences is evaluated over time. There is also a survey targeting the elderly who receive treatment at home or at elderly care facilities—the elderly are asked whether they feel self-sufficient after receiving the services and whether they know about their right to choose their own service provider. There are two websites where Danish residents can get information about the healthcare quality and make an informed choice about hospitals; patients can also access their personal services and patient health records online.

99 percent of all residents in Denmark are registered with a GP under insurance scheme 1. In this case, they have the right to go to a private medical practitioner for specialialist treatment upon referral from the GP. Patients registered in insurance scheme 2 can go to any GP or private medical practitioner as they want, without a referral; however, they may need to pay under the co-payment scheme. The Danish senior healthcare policy aims to promote and extend the independence of senior residents and ensure their self-sufficiency and well-being (Ministry of Health Denmark 2017). The municipalities provide home care and nursing home services free of charge for seniors. The municipalities are also responsible for preventive healthcare measures for senior citizens, including organized social activities, physical training facilities and other volunteer activities. Home visits are also offered to senior residents whenever they need; for seniors over 80 years of age, the visits are offered on an annual basis.

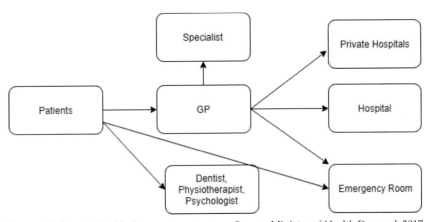

Figure 1.5: Danish Health Care system structure. Source: Ministry of Health Denmark 2017.

As life expectancy in Denmark has increased from 77.9 years in 2005 to 80.7 years in 2016 (WHO 2018), the higher number of patients suffering from dementia is becoming an increasingly serious problem in the aging Danish society. By estimation, about 80,000 people in Denmark suffer from dementia. There have been initiatives and efforts to promote the early detection and precise diagnosis for patients with dementia. The objectives are to make Denmark a dementia patient friendly country where patients can live a dignified life, and to treat patients with dementia according to their needs and values in a coherent way in order to ensure early detection and alignment with the latest technology and research. The other objective is to offer relatives enough support and make sure that they are fully involved in the process (Ministry of Health Danmark 2017). Other than dementia, the number of patients suffering from chronic diseases such as diabetes, arthritis or respiratory diseases has also increased. By estimation, there are a million people in Denmark today suffering from chronic diseases (Ministry of Health Denmark 2017). The two most common chronic diseases in Denmark are diabetes and compulsory obstructive pulmonary disease (COPD). The Danish government has developed an action plan for early detection and treatment of such diseases; it also aims to ensure that all patients receive comprehensive high-quality treatment regardless of their geographical location. The action plan also aims to make sure that the patients are aware of what to expect after the diagnosis, thus promoting patient participation in the treatment and management of the disease, as well as patient empowerment.

The Danish healthcare system is highly digitalized, with digital communication among healthcare providers, systematic data use, and digitalized working procedures. Public hospitals and GPs in Denmark can keep contact with patients over a long period, which makes it possible to perform large-scale data monitoring and analysis. The records stored in the national patient registration and medication databases make it possible to monitor patient compliance. The common data sharing standards enable GPs, hospitals, labs and elderly care facilities to share data with one another (Ministry of Health Denmark 2017). In Denmark, all GPs keep electronic health records and 98 percent of them exchange records electronically. GPs also receive test results from labs electronically, with 99 percent of GPs sending prescriptions to the pharmacies electronically. 97 percent of the patients are referred to the hospital electronically and all patients are referred to specialists and psychologists electronically (Ministry of Health Denmark 2017). There have been efforts to increase the interoperability of the current e-health system; the vision is to support a cohesive system with smooth exchange of information between different service providers.

When it comes to the value-based healthcare system, Denmark has been a forerunner in the field. The linkage between different levels of the patient registration system, and biobanks allow rich data to be used in research and improvement of healthcare services. There has been a national level of reform measures calling on utilizing all the data and creating a better strategy for transparency and monitoring for health outcomes and results. The vision aims to offer better healthcare services through systematic monitoring and benchmarking of results and outcomes, and encouraging better services by setting higher standards; secondly, the results-oriented treatment will focus on providing precise diagnosis and early intervention based on risks predicted by the data for certain segments of the population. Thirdly, the system aims to compare healthcare quality across the healthcare system by comparing key patient information regarding the accessibility and quality of healthcare services, including waiting time and outcomes. The program also aims to promote a management style based on systematic data analysis for health outcome benchmarking whereby the system is rated on the basis of outcomes and efficiency. As a part of the program to establish a more efficient healthcare system, in 2014, a four-year program was launched for health improvement through a better use of data. To achieve this vision, the Health Data program focuses on establishing a new data model and user interface to promote better data use for healthcare professionals, researchers, governments and citizens, on modernizing the IT structure for the national healthcare system, enhancing data quality to improve data validity and reliability, and supporting cross-sectional data cooperation (Ministry of Health Denmark 2017). For combating COPD, Denmark aimed to implement the telemedicine solutions for chronically ill patients at the local (municipalities) and regional levels by 2019, by closely monitoring their key indicators such as oxygen level, heart rate, weight and blood pressure for a couple of weeks; the results were then sent to the local hospitals for analysis, for adjusting medication if needed. The system aims to ease the patient's life by letting them know how exercise and diet can change their heart rate and blood oxygen saturation level. The project reduces the number of hospital admission days and telemedicine seems to be more efficient than ordinary treatment methods.

To summarize, Denmark has led the transformation to the value-based healthcare system with its deeply digitalized healthcare system; data is key for closely monitoring patients throughout the care cycle, establishing health outcome benchmarking, sharing the benchmarking with healthcare professionals across the system and efficiently adjusting treatment for patients. Establishing a high-quality data collection and data sharing platform plays an important role in the value-based healthcare system.

4.1.2 Germany

To facilitate the use of telehealth solutions and improve the efficiency of the telehealth solution approval process, the German government has introduced new regulations and laws. For instance, the digital healthcare act will allow the statutory insurance companies to be able to reimburse patients for home-use IoT medical devices (Stern et al. 2022). The 'fast track' allows low-risk medical devices mainly used by patients at home to get regulatory approval relatively faster. The German Federal Agency for Drugs and Medical Devices (BfArM), by deploying the fast-track approval process for home-use medical devices, strengthens the use of real-world data (real world evidence) to improve data interoperability, privacy and quality of services and solutions offered by medical device providers. Real world evidence is collected through patient reported feedbacks on treatment outcome and user experience. The available funding for digital healthcare solutions has increased exponentially to Euro 3 billion by Bundestag and Euro 1.3 billion by local German governments (Lovell 2019).

The German regulators learned from the FDA to use real world evidence for the approval of medical devices, thereby lowering the cost for tech companies, large or small, to bring digital healthcare devices from the lab to the market. The Digital Healthcare Act also aims to lower the barrier to entry for digital healthcare startups, creating more opportunities for SMEs to enter the market. Needless to say, the Digital Healthcare Act has a strong relationship with how the German healthcare system works and functions.

The German healthcare system is characterized by universal coverage of insurance for a wide range of services (Busse et al. 2014). The system holds a strong solidarity principle whereby treatment is offered regardless of financial status of the patients or the premium paid and the morbidity risks (Ridic et al. 2012, GKV 2019); the principle of benefits ensures benefits without any up-front payment for the insured residents; such residents are also free to choose the services and insurance plan service providers (GKV 2019). The system offers a network of excellent service providers, both private and public. The 'Bismarck' model of social insurance system has extended sick benefits to all low wage workers since 1871. After the reunification of Germany, all 16 provinces ("Laender") have had an independent healthcare policy to a large extent.

All citizens in Germany are now required by law to have health insurance, with more than 90 percent of the population covered by the state statutory insurance system. The rest 10 percent are covered by private insurance or government schemes for students, police force or special assistance. Only those with an annual income of more than 576,000 Euros are entitled to choose private insurance. The state insurance funds are formed through contributions from both the employers and employees; the uniform contribution rate for all insured was introduced on

1st January, 2009. A tool to decide the supplementary premiums for individual insurance funds was also introduced in 2009. Private insurance, however, functions based on the equivalence principle, where the premiums are determined by the benefits and risks agreed upon; in this case, higher premium plans usually cover more services and offer a higher level of reimbursement for treatment such as dental treatment costs.

Statutory health insurance funds have the right for self-governance by the elected members of employers and the insured members. The self-governance insurance funds, through negotiations with healthcare providers, shape a large part of the healthcare system in Germany. The National Association of Statutory Health Insurance Physicians, the National Association of Statutory Health Insurance Dentists, the German Hospital Federation, and the National Association of Statutory Health Insurance Funds form the Federal Joint Committee. The Committee decides on the benefits included in the statutory health insurance (GKV 2019). The benefits include outpatient care provided by doctors working in private practices, inpatient care, home care, preventive services, visits to spas, etc. The contracts between the statutory funds professional organizations and the regional statutory health insurance funds regulate inpatient care (Obermann 2013). The main actors in the German system are thus the professinoal physician organizations such as the National Association of Statutory Health Insurance Physicians, rather than insurance companies or even individual physicians.

Despite the high degree of autonomy for patients and professional autonomies of doctors and patients, government intervention in healthcare reforms is extensive. The federal government in Germany passes legislation for healthcare policies and jurisdiction. The individual state is responsible for hospital planning, managing state hospitals, and supervising statutory insurance funds and doctor associations. Local governments manage the local hospitals and public health programs (Ridic et al. 2012). The rising healthcare cost and the amounting sick funds leads to healthcare reform aiming at higher utilization rate and lower utilization cost. The government is deeply involved in cost cutting reforms. There have been several legislations and acts since the 1970s, aiming at cost control and healthcare quality control, targeting high and rising costs,. Dealing with overcapacity and low reimbursement levels leading to the over utilization of services and costs with no improvement in healthcare outcomes. It is noticeable that, among the three chosen countries, Germany has incurred the highest amount of healthcare costs in real terms as well relative to GDP, as depicted in Figure 1.6.

For instance, in Germany, 300 hospitals cover the 18 million people in North-Westphalia, while only 70 hospitals cover the same population in the Netherlands. The current reform, however, has been focusing on shutting down hospitals and reducing the number of doctors rather than

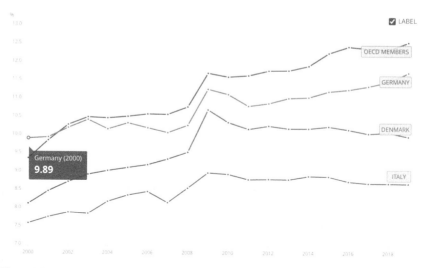

Figure 1.6: Healthcare expenditures as percentage of GDP for the selected EU countries and OECD member states. Source: World Bank 2018.

optimizing the existing resources (EIU 2016). In general, there is a large variation in inpatient and outpatient treatment. For outpatient treatment, primary level doctors do not have such a strong role as in Denmark. Patients can visit a specialist without referral from a family doctor; the primary level healthcare does not play a strong role in coordinating the care process as the healthcare system itself is quite fragmented. Different service providers vary greatly in terms of service quality.

At present, the Statutory Health Insurance System encounters several challenges brought by the changing structure of the population, the first major challenge being the declining amount of contribution to the insurance system, with high-income groups migrating to the private insurance system, along with a drop in the population liable for compulsory insurance; secondly, the aging population is driving up healthcare costs; thirdly, the income portfolio of households has become less wage dependent. With the advancement in technology, the inefficient structure of the healthcare system and the fast-rising cost for certain sectors are driving up costs for the healthcare system. The following chart (Figure 1.7) suggests that hospitals incur the highest amount of costs in the German healthcare system. Data suggests that from 2003 to 2013, inpatient care accounted for about 32 percent of public health expenditures; the corresponding number for outpatient care was about 23 percent, and that for pharmaceuticals and other medical non-durables was about 15 percent (European Commission 2016).

There have been several reforms targeting the rising healthcare costs in Germany, as well as an initiative to develop an e-health system. These

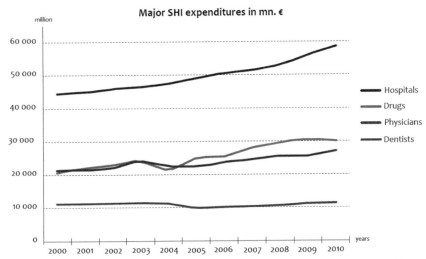

Figure 1.7: Major Statutory Insurance spending in Germany (from top to bottom: Hospitals, Drugs, Physicians, Dentists), 2000–2010. Source: Obermann 2013.

reforms focus on changing the traditional holistic approach, whereby the health system focuses on collective goals which are, to a large extent, monetary. Quality evaluation is based on economical results, with individual patients categorized into a specific group. The management is mainly based on a top-down approach. The new scheme intends to determine the appropriate resources depending on the individual patient-physician relationship. Healthcare expenditures are made based on the perceived responsibility of the individual patient, aimed at an optimal result with appropriate costs (Obermann 2013).

E-Health in Germany is supposed to contribute to better medical care provision, improve communication among all parties involved and ensure high efficiency in the healthcare process (European Commission 2016). Obermann et al. identified the following stakeholders for the implementation of E-health solutions, taking into account all the actors in the public healthcare system, including insurance companies, physicians, dentists, pharmacists and hospitals. An important project for expanding E-Health in Germany is the introduction of an E-Health card; 2018 onwards, patients can choose to store their emergency medical information, electronic medication plan and electronic health records therein (Obermann et al. 2013, European Commission 2016, AOK 2018). Gematik, the system constructor, ensures that the electronic health records contain information regarding discharge letters, medication history, emergency medical data maintained, and examination images by healthcare professionals. From 2019 onwards, patients have the option to log their own data such as heart rate or blood glucose level in the e-health

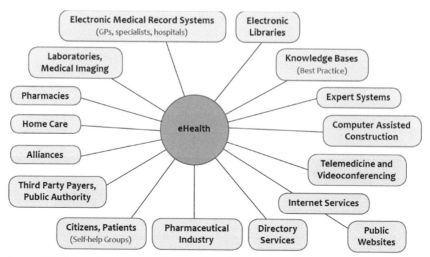

Figure 1.8: Stakeholders in the German Healthcare System for implementing e-Health solutions. Source: Obermann et al. 2013.

folder on the card (European Commission 2016, Obermann et al. 2013, AOK 2018). 75 percent of the population, with 93 percent of the German population suffering from chronic disease conditions, are willing to send data on their vital signs to the doctor (Bitkom 2016, Rinbeat 2018). One-third of the German population are already using wearables—not just for monitoring sports activities, but also for monitoring vital signs. This suggests that there is a large demand side request for telehealth solutions to get implemented (Rinbeat 2018).

To summarize, the smart healthcare solutions in Germany are closely linked to empowering the patients and changing the traditional top-down management approach to distribute healthcare resources in the German healthcare system. The top-down approach has led to the over-utilization of medical resources. Telehealth solutions can potentially link doctors, hospitals, pharmacies, and insurers, thus allowing the streamlining of administrative tasks and cost reduction. Access to patient's medical history and previous examinations can make diagnosis more precise, treatment more targeted and recoveries faster. Patients have the right to decide upon the data to be shared with the doctors on their health folders (Obermann et al. 2013). However, there are other challenges for the German health system, such as enhancing the capacity and use of GPs in the healthcare system, and promoting healthcare spending efficiency. Evidence shows that using wearables can reduce hospital costs by 16% over 5 years (Rapid Value Solutions 2020). Given that hospital costs contribute to the largest part of German healthcare costs, and are rising at a faster pace than physician costs and drug costs, the implementation of smart health solutions is worth studying in order to reduce hospital stay and promote patient self-care, thereby reducing healthcare costs.

4.1.3 Italy

Italy is an interesting case for this study, given the north-south differences in the economic development level and thus diversification of health service quality. The Italian health system is decentralized to a large extent, with most regions managing the organization of healthcare; the national level has a limited amount of power (Cicchetti and Gasbarrini 2016). The state has full control of the core benefit package, but evidence shows that the service quality varies greatly across regions (Cicchetti and Gasbarrini 2016). The Italian national health service (NHS) is organized at national, regional and local levels (European Commission 2017). The Ministry of Health takes a stewardship role in the healthcare system, determining the core benefits packages and allocating budgets to regions (Cicchetti and Gasbarrini 2016). The regional healthcare authority is responsible for delivering community healthcare, primary care, and specialist care with physicians or public hospitals or approved private practices (Cicchetti and Gasbarrini 2016).

The decentralization is based on the idea that localization can be the best option to meet local healthcare needs (Cicchetti and Gasbarrini 2016); Due to decentralization policies, different regions show a large divergence in terms of public resources available for healthcare because of differences in economic development, as well as regional infrastructure. Traditionally, northern Italy, which is more industrialized than Southern Italy, have better healthcare services for residents. For instance, the northern and the central regions have higher healthcare capacity, more advanced technology and better perceived quality of care compared to the southern region. In the end, patients flow from the south to the north for better treatment. Almost 30,000 patients leave the area of Campania, Calabria and Sicily per year for better quality of care in the north (Ministry of Health Italy 2011). The healthcare spending in southern regions, such as Campania, was over 40 percent less than the national average spending on health in 2016 (Ministero dell'Economia e delle Finanze 2020, Istat 2017). After the financial crisis of 2008, there have been calls to recentralize the healthcare system. As a result, half of the regions report deficit in the health sector (European Commission 2017).

Compared to other European countries, the out-of-pocket spending ratio for Italian residents is high, even though the Italian healthcare system provides universal healthcare coverage to those with a residence permit (European Commission 2017). 23 percent of healthcare expenditures in Italy are paid out of pocket, compared to EU average of 15 percent in 2015. Primary and inpatient care are free, while co-payment is applicable

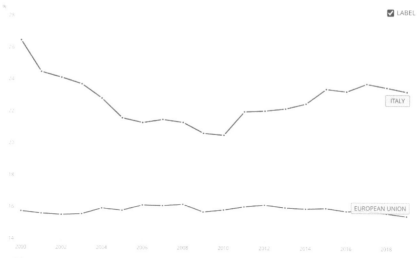

Figure 1.9: Out of pocket payment expenditure in current health expenditures (in %).
Source: World Bank 2022.

for specialist visits with GP referral, on diagnostic procedures and medications (European Commission 2017). Because of the austerity of measures after the financial crisis, patient out of pocket payment ratio has been rising. The system has proved to be pro-rich for specialist visits, diagnostic procedures and basic medical tests, and pro-poor for primary care. A possible reason for this is that for those in the higher socio-economic class, health literacy level is high, while free primary care, long waiting times and low-quality services in the southern region are driving the higher income population to private healthcare services (European Commission 2017). Private healthcare expenditures in rich regions such as Bolzano, Lombardia and Valle de Aosta is about two times that of the poorer regions of Campania and Calabria (European Commission 2017).

Aging has become an issue threatening the financial stability of the healthcare system. In Europe, Italy has become one of the oldest populations, with an age dependency ratio (old) of 37 percent in 2017, as shown in Figure 1.10. In 2016, public health expenditures accounted for 75 percent of all healthcare costs (equivalent to 6.7 percent of GDP), while total healthcare expenditures accounted for 8.9 percent of GDP (The Italian National Institute of Statistics 2017). Hospitals costs contributed to 45.5 percent of total healthcare expenditures, while the corresponding number for outpatient care was 22.4 percent (The Italian National Institute of Statistics 2017). In the long term, as aging intensifies, the population with one or multiple chronic conditions will also rise. This will bring a burden to the healthcare system.

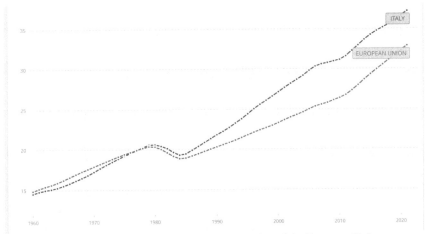

Figure 1.10: Age dependency ratio (Old) in Italy and the European Union.
Source: World Bank 2022.

When it comes to the electronic healthcare system, Italy lags behind in the exchange of data via internet and offering e-prescriptions. Only 3 percent of GPs send administrative data to other care providers via internet, and only 1 percent of the GPs use e-prescriptions (Ferré et al. 2014). In some regions, pilot programs have been running for implanting E-health solutions. The collection of administrative data and patient history on electronic patient records have been deployed in Lombardy and Emilia Romagna (European Commission 2019b). In Emilia Romagna, the project 'SOLE – Online Healthcare' has aimed to develop an integrated network for hospitals and physicians, including the function of offering e-prescriptions for patients. The 'Renewing Health' EU-funded project in the Veneto region targeted the adoption of telemedicine services for chronically sick patients with COPD and diabetes (Renew Health 2009).

However, wide adaptation of telehealth solutions remains a challenge. The delivery of healthcare and decision-making vary on the regional level, depending on quality, financial resources and priorities. There are two areas to which telehealth solution can contribute in Italy: (a) Telehealth can facilitate establishing a national level evaluation of healthcare quality, therefore aiding a more efficient use of public spending on healthcare, and (b) Telehealth solutions can help promote access to healthcare, reduce waiting times, and improve healthcare quality in the primary level of care. This requires long-term dedication to the Telehealth initiative, where Italy stands at a preliminary level. Currently, there are two national level policies aimed at promoting smart health. The first initiative is the implementation of the New Health Care Information System (NSIS, Nuovo Sistema Informativo Sanitario), an information system to supervise all primary level healthcare services. The second initiative is E-Government

Plan 2012, which aims at developing online solutions and EHR systems, and offering an e-prescription system for patients. The impacts of these initiatives are yet to be evaluated (Ferré et al. 2014).

4.2 Challenges for the Chinese healthcare system and implications for telehealth

4.2.1 Challenges for the healthcare system in China

The majority of inpatient and outpatient care in China are performed by public hospitals, in which level III and II hospitals account for the most patient visits (National Health Commission 2018). The designated family doctor services, which fall under the jurisdiction of urban community care centers and rural village clinics, play a limited role in the diagnosis and treatment process (SCMP 2019).

For outpatient care in China, the key problem is the long patient waiting times (Sun et al. 2017). Patients prefer public hospitals for better facilities, services, and qualification of doctors (Mckinsey 2010). In a Chinese hospital, patients wait in several queues for registration, payment for treatment and getting prescribed medicines separately. In case doctors find it necessary to perform diagnostic tests such as MRI scan or X-Ray imaging, patients need to wait in the queue to pay for these diagnostic tests. After payment, patients can take the receipt to the nurse station and get a number for the test. After waiting for the number to be called and conducting of tests, patients can take the receipt back to the doctor for interpretation. If lucky, the process can take up to half a day or one day with patients going home with a large bag of medication prescribed by the doctors (SCMP 2019). In all probability, these public hospitals will only suffer from the aging population and intense patient-doctor relationships.

The growing demand from the aging population for high-quality medical services has prompted tech companies to work together with healthcare institutions to develop new healthcare models. For instance, Alihealth works with hospitals to improve diagnosis accuracy (Alihealth 2019); patients can now register for doctor appointments via the Alipay app. Tencent works to develop a family doctor service app and has acquired clinics, offering reliable primary care services to patients; it has also invested in the Wedoctor platform, making it possible for patients to register for primary care services over Wechat (Reuters 2018). Huawei has worked with hospitals to develop heart rate monitoring wearables (Huawei 2019), whereas Xiaomi developed Mi Band to monitor the user's sleep patterns (Xiaomi 2019).

For inpatient care in China, hospitalization is expensive, particularly for patients with no insurance coverage in rural and urban areas, who need to pay out of pockets. Most hospitals are crowded for both inpatient and outpatient care for the same reason that most medical resources, such as qualified doctors, advanced medical devices and useful medication can only be found in 'Tertiary Level Hospitals'. Both rural and urban patients in China have the incentive to go to level three hospitals for diseases ranging from flu to cancer. This has resulted in overused and burdensome hospitals and insufficient use of primary care centers and other primary care facilities. Table 1.4.1 describes the utilization rate of all levels of hospitals in China in Year 2021.

Compared to 2017 (see Table 1.4.2), in 2021, the number of hospitals grew in China, while the bed utilization rate as well as the number of outpatients and inpatients dropped for level II and level I hospitals (not for Level III hospitals). Considering the impact of Covid and the zero-Covid policy, it is evident that access to healthcare services has deteriorated during the Covid time, while healthcare cost has risen. Hospitals, which are financially independent from one another, pass on the cost to the patients. Compared to Level III hospitals, the number of outpatients and inpatients for level II and level I hospitals has dropped significantly; the outpatient number for level III hospitals also declined by 0.62 billion.

It is noticeable that the number of Level III hospitals and the number of outpatient treatment are not proportional. This explains the long waiting times at hospitals and the patients' complaints of poor services received, including seeing doctors for merely 2 minutes after a waiting time of half an hour (Lim 2014).

Table 1.4.1: Hospital bed utilization rate in China in 2021. Source: National Health Commission 2022.

Year 2021	Number	Beds	Bed Utilization (%)	Number of outpatients (Billions)	Number of Inpatients (Millions)
Level III Hospital	3275	> 500	85.3	1.11	112.46
Level II Hospitals	10848	100–499	71.1	0.561	68.9
Level I Hospitals	12649	20-99	52.1	0.108	11.2

Table 1.4.2: Hospital bed utilization rate in China in 2017. Source: National Health Commission 2018.

Year 2017	Number	Beds	Bed Utilization (%)	Number of outpatients (Billions)	Number of Inpatients (Millions)
Level III Hospital	2340	> 500	98.6	1.73	84
Level II Hospitals	8422	100–499	84	1.27	80
Level I Hospitals	10050	20–99	57.5	0.22	11.7

The Chinese government has taken measures to ensure universal healthcare coverage. The Employee Basic Medical Insurance (EBMI) covers all state-owned and privately owned firm employees on a mandatory basis. The Resident Basic Medical Insurance (RBMI) covers the rest of the residents, such as children, the elderly, students and the unemployed population; it operates on a voluntary basis. By the end of 2018, roughly 95 percent of the population had been covered by health insurance. However, medical expenses still account for a major expense for the Chinese. The reason is that the scope and the extent of cover by the EBMI and the RBMI remain limited. The following form describes the coverage scope of EBMI, RBMI and New Cooperative Basic Medical Insurance (NCBMI). With the ongoing medical reforms, the NCBMI has been integrated with URBMI in certain provinces to form the RBMI.

The scope and the extent of refund for insurance scheme coverage varies with the location of the insured subject; for instance, the percent of contribution paid by employers for EBME varies across cities. Take Wuxi-Jiangsu Province, Shaoxing-Zhejiang Province, Nanchang-Jiangxi Province and Guiyang-Guizhou Province for example. In these provinces, the payment cap for outpatient treatment ranges from RMB 50 to RMB 900 for regular diseases, and that for catastrophic diseases ranges from RMB 80,000 to RMB 110,000, with insurance coverage ratios ranging from 30% to 100%. Furthermore, the payment cap for inpatient treatment ranges from RMB 35,000 to RMB 100,000, with insurance coverage ratios ranging from 30% to 100%. As hospital costs rose while the percentage of out-of-pocket payment also increased significantly, the healthcare system has become more pro-rich, thus rendering rural residents subject to a lack of reliable healthcare resources (Mckinsey 2010).

This has created a room for smart health solutions operating in pharmaceutical e-commerce, Automatic machines for pharmaceuticals, tracking for authentic medicine, and consumer health. For one of the largest smart health businesses in the Chinese market—Alihealth, the online-offline medication and service sale has shown a year-on-year growth of 296.8% and 275.5% in financial years 2018 and 2019 respectively (Alihealth 2019).

In China, patients usually pay for a small government-capped fee for doctor visits, yet hospitals are free to charge for check-ups, prescribed drugs and specialized treatments (EIU 2016). The government funds 10 percent of healthcare costs at public hospitals; thus, for hospitals, prescribing medication and diagnostic tests make up for the rest of the 90 percent of the cost and profits (EIU 2016). This leads to over-prescription of medication, overuse of image examinations and other services which profit hospitals and clinics. This financial model also leads to the under use of primary level care resources.

Table 1.5: Public Health Insurance Coverage in China in 2016. Source: National Health Commission 2018.

Domain	Category	Eligibility	Covered Population (million) by 2021	Coverage Cap (RMB), Ratio
Public Health Insurance in China	Urban Employee Basic Medical Insurance (UEBMI)	Mandatory Basic Insurance, covered by employers and employees, annual premium $100–250	316.81	Varies by region, Outpatient coverage: 10-500,000, Co-payment: Average 60%
	Resident Basic Medical Insurance (RBMI)	Voluntary Basic Insurance, for urban residents not eligible for UEBMI, Funded by government and individual, annual premium $20–100	897.36	Varies by region, Outpatient coverage: 0–500, Co-payment: Average 50%
	New Rural Cooperative Medical Insurance (NRCMS)	Voluntary Basic Insurance, for rural residents, Funded by government and individual, annual premium $20–50	130	N/A, because of ongoing medical reforms, NRCMS is going to be integrated into RBMI

Table 1.6: Healthcare cost structure in China in 2021. Source: National Health Commission 2022.

Healthcare Cost in China	Healthcare cost as percentage of GDP (2021)	6.5%
	Government Public Expenditures	27.4%
	Social Healthcare Expenditures	44.9%
	Personal Healthcare Expenditures	27.7%
	Annual average medical cost per person	USD 739.88

Meanwhile, with an aging population, the Chinese healthcare system is facing a different challenge than before. Life expectancy at birth in China has reached 78.3 years overall (National Health Commission 2022). More healthcare issues are now arising from chronic conditions instead of acute conditions and injuries. At the same time, age dependency ratio (old) has risen to 43% in 2021, suggesting that aging has accelerated in China.

Table 1.7 describes the leading causes of death in China. Chronic diseases have become the leading cause of death in China, such deaths reaching up to 86.6% in 2012 (National Health and Family Planning Commission 2015). Cardiovascular diseases have become the primary

Table 1.7: Health Status of the Chinese population and composition of death rates. Source: National Health Commission 2021.

Indicators	China	
	Rural China (2018)	Urban China (2018)
Life Expectancy (years)	76.8	78.9
Infant mortality (‰)	7.3	3.6
Under-five mortality (‰)	10.2	4.4
Maternal mortality (1/100,000)	19.9	15.5
Leading cause of death: Composition of total death rates (1st to 10th) (2018)	Cardiovascular disease/26.89%	Cardiovascular disease /26.41%
	Cerebrovascular disease/24.31%	Malignant Neoplasm (Cancer)/22.19%
	Cancer/19.1%	Cerebrovascular disease/21.45%
	Respiratory disease/11.35%	Respiratory disease/10.27%
	Injury and poisoning/5.86%	Injury and poisoning/4.84%
	Endocrine nutrition metabolic disease/3.16%	Endocrine nutrition metabolic disease/4.05%
	Digestive system disease/1.78%	Digestive system disease/2.08%
	Neural System Disease/1.43%	Perinatal disease/1.59%
	Infectious and parasitic diseases/1.03%	Neural System Disease/1.08%
	Infectious and parasitic Disease/0.74%	Infectious and parasitic diseases/0.67%

cause of death in China, followed by cerebrovascular diseases, respiratory diseases, perinatal diseases, digestive diseases and nutrition metabolic diseases (National Health Commission 2018).

Among the factors leading to chronic diseases, environmental risks, behavioral risks and metabolic risks are contributing to the loss of life in China (Zhou et al. 2019). Behavioral risks include dietary risks, under-nutrition, unsafe sex, physical inactivity, sexual abuse, and violence, and tobacco, alcohol and drug use. Environmental risks include air pollution, water pollution, sanitation issues, handwashing, occupational risks, and other factors. Metabolic risks include high blood pressure, high body-mass index, high plasma glucose, high cholesterol levels, low glomerular filtration, and low bone mineral density (Zhou et al. 2019).

For the Chinese population, hypertension, obesity, over-intake of sodium, physical inactivity, nutrition imbalance, high cholesterol and glucose levels, diabetes, unsaturated fat, lack of intake of vegetables and fruits, and use of tobacco and alcohol have been identified as the key risk factors (National Disease Control Center of China 2013). Among these risk factors, hypertension stands out; a survey from 2012 to 2015 involving 451,755 residents from 31 provinces revealed that 23.2% (approximately 244.5 million) of the Chinese adult population suffered from hypertension and 41.3% (about 435.5 million) had pre-HTN according to the Chinese guidelines; Among individuals with hypertension, 40.7% were taking the prescribed medications, 46.9% were aware of their conditions and 15.3% had their conditions under control (Wang et al. 2018b). Prevalence of hypertension according to the American College of Cardiology/American Heart Association guidelines was twice as high as that based on the Chinese guidelines, reaching 46.4%, with control rate for the condition falling to 3.0% (Wang et al. 2018b).

With aging intensifying in China, more general practitioners need to perform care at home and deal with routine consultations (The Economist 2017). The reforms in China have focused on two targets to direct patients to the primary care level: the first goal is to make healthcare cheaper for patients, and the second one is to revive primary healthcare by diverting patients from hospitals to local clinics. The government had pledged to establish 7,000 more urban community health centers (CHCs) and small hospitals with higher standards in rural areas by 2011. However, the reform did not successfully divert patients to primary level care for several reasons.

The first reason is that no official gate-keeper system exists on the primary level of healthcare in China, meaning that there is a lack of professionally trained general practitioners in the country. Currently, there are about 60,000 licensed general practitioners (GPs), accounting for just 3.5% of all doctors in China; this leaves 650 million Chinese without access to a GP (Wang et al. 2018a). Most medical schools in China do not offer general practitioner training, with students choosing to become specialists since the third year of medical school. Even under the 5+3 training model aimed at training general practitioners in China after students complete their bachelor's degree and the subsequent three years of clinical practice, only a small portion of those students choose to become general practitioners rather than pursuing a higher degree. In the study of 2017, students who choose to become a GP were either female or came from rural areas or from families with a low monthly income (in the range of RMB 4,000, equivalent to about 521 euros) (Wang et al. 2018a).

Secondly, there is less incentive for medical students to become general practitioners; compared to primary care institutions (such as community healthcare centers) and secondary hospitals, tertiary hospitals offer better payments, more career advancement opportunities and higher social status to doctors. A doctor working at a hospital earns an annual income of RMB 80,000 on an average (equivalent to 10,397 Euro), with well-trained specialists usually having a huge opportunity to receive 'red envelopes' as thank-you notes from desperate patients. A GP in China usually earns about RMB 50,000 (equivalent to 6498 Euro) (The Economist 2017). This makes career opportunities at tertiary hospitals more attractive for highly educated and well-trained doctors.

Thirdly, most doctors working at primary care facilities are not well qualified. According to the Guideline for Establishing the General Practitioner System promulgated by the Chinese government, less than 23 percent doctors at community healthcare centers in rural areas have a bachelor's degree; less than 4 percent of doctors in community health centers have a senior title, and less than 56.7 percent of country-side doctors are qualified to apply for the registered doctor license in China.

Lastly, most Chinese patients disdain primary care, even with treatment received from fully qualified GPs. This is because most GPs are not authorized to prescribe a wide range of drugs as hospital doctors are; besides this, most Chinese do not trust GPs as this reminds them of the bare foot doctors[2] age (The Economist 2017). Most Chinese patients feel comfortable seeing a university educated doctor at a hospital with modern facilities (The Economist 2017). People also have less incentives to consult a GP; the basic government insurance coverage leaves patients to pay anything between 30 percent and 40 percent of their outpatient treatment costs wherever the treatment takes place. Seeing a GP risks being referred to a specialist later, resulting in higher costs. Therefore, most Chinese prefer getting treated at the hospital and seeing a specialist directly (The Economist 2017). Until now, establishing a hierarchical healthcare system has remained challenging in China.

With the pandemic crowding out resources from the healthcare sector to pandemic prevention and control, primary healthcare institutions suffered the greatest blow. The bed utilization rate dropped from 57.5% in 2017 to 52.1% in 2020 and 2021, along with the number of days of patients' stay at level I hospitals. The financial implications for Level I hospitals are considerable as well. Resources are mostly pulled from public finance to support pandemic control methods when public finance remains the major source of financing primary level of healthcare in China.

Meanwhile, because of the decentralized healthcare system, healthcare quality varies greatly with geography in China (EIU 2016). Differences in healthcare quality arise from the fact that different provinces operate

---

[2] In Maozedong era, "Barefoot Doctors", in Chinese "赤脚医生" are doctors who serve in the rural area (villages, towns) with basic medical education.

independently to provide healthcare, with healthcare quality higher in coastal areas and big cities, and lagging in inland regions (EIU 2016). For example, the number of GP per 1000 people is twice as high in coastal areas compared to western and central China.

The healthcare reform has been initiated by the State Council in China, merging the previously fragmented healthcare service regulators into a single entity: National Health Commission (NHC) in China (National Health Commission 2019). The NHC has promulgated the 'Internet + Healthcare' initiative (State Council 2018, National Health Commission, 2019). This prompted tech companies to partner with hospitals for infrastructure development for cloud storage and processing of data, and develop new business models in healthcare. This also gave insurance companies the opportunity to work with tech companies for online medical insurance sale and compensation schemes. In some pilot cities, it is possible to use the basic medical insurance schemes to pay for services and medical products and reimburse online. The rural cooperative medical insurance scheme (NRCMI) has been integrated with URBMI in the medical reform to reduce the medical care cost for rural citizens and migrant workers (State Council 2016). So far, the insurance scheme has been able to cover over 90% of the population in China.

4.2.2 Implications for telehealth

The COVID-19 pandemic gave a new perspective on the demands for telehealth solutions in different healthcare systems. For instance, in China, digital healthcare tools are used to monitor large-scale population movement and whether people have close contacts with patients of COVID-19 symptoms and positive cases. However, the Chinese system seems to overreach to the point where the GPS location of people are tracked regarding whether they have been to local government assessed low-risk, middle-level-risk or high-risk areas. Often millions of people are quarantined for a few positive cases in the region. The access to healthcare services has deteriorated, whereas the cost has risen for receiving healthcare services. In other economies, however, COVID has crowded out medical resources as well, leaving patients to face long waiting times and dropping healthcare quality. How telehealth solutions interact with pandemic prevention and control will be discussed in Chapters 3 and 4.

There have been calls to promote the value-based healthcare system, whereby the outcome of treatment relative to the cost becomes the primary index, catering for evaluating healthcare quality. The model aims to control rising healthcare costs, enhance healthcare quality, take care of patient complaints on excess use of medication and diagnostic tests, and reduce the opportunity cost on overutilization of existing resources, diverting them towards investments in new drugs, clinical pathways and technology.

In the pilot program promoted by the World Bank at the People's Hospital of Yiyang in Henan Province, the hospital introduces clinical pathways for certain medical conditions and tries to move away from the payment-for-service model to case-based payment, thus diverting incentive for overcharge on drugs and tests. The evaluation criteria for payment are based on whether the patient has successfully completed treatment following prescribed evidence-based clinical pathways for the condition or disease (EIU 2016). The doctor's income depends on the number of patients successfully treated under the clinical pathway. If the pathway is not followed due to errors or inaction, there will be financial penalty imposed on the care team. If there is successful completion of pathways at the time of discharge, there might be a bonus for the team (EIU 2016).

Data was collected on several indexes to make sure that the clinical pathways were followed, including the number of patients enrolled in the clinical pathway protocols, readmission rates at 14 and 31 days, hospital acquired infections, drug outlays, health insurance reimbursement amount, out-of-pocket spending and patient satisfaction rates. The evaluation of whether the clinical pathways were followed was done through a sophisticated IT system to monitor patient condition and whether the protocol was followed in real time (EIU 2016). The People's Hospital in Yiyang claimed shortening of the length of hospital stays by a full day and reported an 8.7 percent increase in revenue, with a better reputation attracting further more patients. Unnecessary treatment dropped by 20 percent, and communication between patients and medical staff improved as well (EIU 2016).

Telehealth solutions are capable of monitoring key health indicators such as heart rate, sleep quality, weight, etc. This aids in the implementation of the three foundational principles of value-based care delivery: (a) systematic measurement of the health outcomes that matter to patients and the costs required to deliver those outcomes across the full cycle of care, (b) identification of clearly defined population segments and the specific health outcomes and costs associated with those segments, and (c) development of customized segment-specific interventions to improve value for each population segment (World Economic Forum and Boston Consulting Group 2017). Telehealth solutions also provide a good opportunity to make sure that the care team follows the same protocol to treat patients, and enables the routine collection of medical data to share and analyze data on healthcare outcomes, e.g., survival rate for patents with pro-state cancer (Helissey et al. 2023) analyze population segment-wise medical costs for patients, given that the cloud platform can collect data from multiple systems including healthcare institutions and public and private insurance systems. By collecting data on detailed health outcomes, it is possible to identify the best treatment practices and reduce inefficiency (Philips 2019).

For lack of utilization of primary level of healthcare, telehealth service providers have developed some innovative solutions through which internet-based hospitals in China are providing online patient consultation, online prescription and long-distance patient treatment services (Wuzhen Internet Hospital 2019, Ping An Good Doctor 2019). Ping An has developed a '1-minute clinical box' for patients to consult doctors in the box and get online prescriptions through an automatic machine outside. The Ping An '1-minute clinical box' service is supposed to be free, with patients only paying for medication. So far, the boxes have become available at High Way service stations in Nanjing (Sohu Finance 2019). In the future, Ping An has planned to provide the service at airports, railway stations, shopping malls and Ping An Good Doctor network clinics and hospitals (Sohu Finance 2019).

5. Conclusion

The current discrepancies in the healthcare system and the unsatisfactory patient experiences have made room for the smart healthcare industry to mend the gaps. For instance, to tackle the problem of inter-regional and international patient data transfer, there have been standards developed to facilitate such data transfer in the EU. Online appointment system and mobile phone apps have been developed to help patients book appointments at hospitals and reduce waiting times. Whereas chatbots and apps can deal with stress, wearables can monitor key bio-metric signals of patients and facilitate chronic disease management. Sensors at home can detect the living patterns of the elderly—such as sleep, eating and falls. The latest technological developments offer hopes for the elderly to live independently at home. This also provides opportunities to governments to further digitalize the healthcare system, improve data transparency for patients and improve health equity for patients in remote areas.

Current healthcare policies in Europe and China focus on promoting access, universal healthcare, cost control and improving healthcare quality (Terry 2016). These pillars reflect the impossible trinity of access versus cost versus quality (Terry 2016). The value-based healthcare system proposes a solution where the healthcare service providers are rewarded on the basis of health outcomes rather than the number of services provided. Transforming into such a system demands the healthcare system to perform systematic evaluation at each stage of healthcare, which is only achievable through comprehensive data collection and analysis.

Europe, like China, varies in terms of digitalization for the healthcare system. In both the healthcare systems, the public sector plays an important role in stewardship and guidance. Public spending also accounts for a large part of healthcare expenditures. Out-of-pocket payment ratios are higher in China, at 28.8 percent, compared to the EU average of 15 percent. Aging and the management of chronic disease have proved to be a common challenge for the healthcare systems in China and the EU. Establishing standards for benchmarking the best clinical pathways and practices are again challenges in both China and EU. For this, full digitalization of the healthcare system is necessary, establishing a national system with patient databases, a system whereby different levels of healthcare institutions and professionals can exchange data, patients can store vital health data and medical history, and share them with doctors, healthcare service providers, pharmacies and insurance companies (public or private). Improving data quality and system interoperability and promoting the use of eHealth solutions are common challenges in the EU and China towards the realization Telehealth solutions for dealing with challenges associated with the aging society.

China and Europe differ in many aspects, but they share the common challenge of aging and rising costs in the healthcare sector, brought by chronic diseases such as diabetes, COPD, obesity and hypertension. Using smart health solutions can potentially help establish a value-based healthcare system to maximize health outcome and minimize costs. The key for transformation into a value-based healthcare system is systematic evaluation at and of each stage of health care, with thorough data collection, analysis and subsequent benchmarking of best practices. With such a system in place, it is possible to practise telemedicine, perform long-distance monitoring of patients and empower patients for self-care. It is also possible for telehealth solutions to fill in the gap where primary care and the gatekeeper function are weak. Improving the use of internet for exchanging data, reducing cost and improving system interoperability is a challenge for the desired success. In the next chapter, a detailed analysis of relevant stakeholders, their roles, powers and attitudes towards the use of smart healthcare solutions is performed.

In the next three chapters, the potential of use of telehealth solutions, challenges towards their implementation, and the role and power of each stakeholder have been elaborated further.

List of Abbreviations

AF	Atrial Fibrillation
AI	Artificial Intelligence
API	Application programming interface

B2B	Business to Business
CD	Compact Disk
CD-ROM	Compact Disc Read-Only Memory
CHCs	Community health centers
COPD	Chronic Obstructive Pulmonary disease
ECG	Electrocardiography
EEA	European Economic Area
eHealth/ E-Health	The use of information and communication technologies (ICT) for health
EHRs	Electronic Health Records
EIU	Economist Intelligence Unit
e-journals	Patient records from hospitals
EU	European Union
Eurostat	The EU Open Data Portal
FDA	The US Food and Drug Administration
FHIR	Fast Healthcare Interoperability Resources
GDP	Gross domestic product
GKV	Gesetzliche Krankenversicherung
GP	General Practitioner
HALE	Healthy life expectancy
HEART	Health related Activity Recognition system based on IoT an interdisciplinary training program for young researchers
HIS	Hospital Information System
Telehealth	Internet of Things in Healthcare
IOS	A mobile operating system created and developed by Apple Inc
IoT	Internet of Things
IoT	Internet of Things
IT	Information Technology
JAMA	The Journal of the American Medical Association
MRI	Magnetic resonance imaging
NRCMS	New Rural Cooperative Medical System
NHC	National Health Commission
O2O	Online to Offline
OECD	The Organisation for Economic Co-operation and Development
PPG	Photoplethymographic
Q1	Quarter I
RMB	The official currency of the People's Republic of China

SHI	Social Health Insurance
EBMI	Employee Basic Medical Insurance
UI	User Interface
RBMI	Resident Basic Medical Insurance
US	The United States
VO2	The maximum rate of oxygen consumption measured during incremental exercise
WHO	World Health Organization

References

Accenture. (2018). ARTIFICIAL INTELLIGENCE: Healthcare's New Nervous System. https://www.accenture.com/us-en/insight-artificial-intelligence-future-growth.

Agency for Digital Italy. (2018, May). National Agenda for e-Health. https://www.himss.eu/sites/himsseu/files/Basile_Italian_Vision_Smart_Healthcare_National_Agenda_e-Health.pdf.

Alihealth. (2018.). Online and Offline Integrated Healthcare solution. Retrieved November 28, 2018, from https://www.alihealth.cn/.

Alihealth. (2019). Alihealth Annual Report 2018/19. http://doc.irasia.com/listco/hk/alihealth/annual/2019/ar2019.pdf.

Allen, J. (2007). Photoplethysmography and its application in clinical physiological measurement. Physiological Measurement, 28(3): R1–R39. https://doi.org/10.1088/0967-3334/28/3/r01.

AOK. (2018). Electronic Healthcare Card in Germany. https://www.gemalto.com/govt/customer-cases/germany-second-generation.

Apple. (2018, January). Apple announces effortless solution bringing health records to iPhone. Apple Newsroom. https://www.apple.com/newsroom/2018/01/apple-announces-effortless-solution-bringing-health-records-to-iPhone/.

Apple. (2019). Apple Watch Series 4 - Health; ECG – Apple. Retrieved January 14, 2019, from https://www.apple.com/watch/.

Apple. (2021, March 8). Institutions that support health records on iPhone and iPod touch. Apple Support. https://support.apple.com/en-us/HT208647.

Asia Insights. (2018 March 19). One Billion Users And Counting—What's Behind WeChat's Success? Forbes. https://www.forbes.com/sites/outofasia/2018/03/08/one-billion-users-and-counting-whats-behind-wechats-success/#1842eaee771f.

Bao, L., Han, R., and Wang, Y. (2014). Analysis of human resources deployment and the policies in China. Chinese Journal of Hospital Administration, 3: 197–201. http://caod.oriprobe.com/articles/41213303/Analysis_of_human_resources_deployment_and_the_policies_in_China.htm.

Bates, D. W., and Wells, S. (2012). Personal health records and health care utilization. JAMA, 308(19): 2034. https://doi.org/10.1001/jama.2012.68169.

BfArM. (2019). Digital Health Applications (diga). https://www.bfarm.de/EN/Medical-devices/Tasks/DiGA-and-DiPA/Digital-Health-Applications/_node.html.

Bitkom, e. V. (2016). Telemedizin trifft auf großes Interesse. https://www.bitkom.org/Presse/Presseinformation/Telemedizin-trifft-auf-grosses-Interesse.html.

BLED. (2010). Proceedings, Yu, Y., Wilfred, H., Juergen, S., and Nilmini, W. (2010). Evaluation of e-Health in China. BLED 2010 Proceedings|BLED Proceedings|Association for Information Systems. https://aisel.aisnet.org/bled2010/.

Bloomberg. (2018a, April 19). Advice for the Good Doctor: Stop Smoking Cash, https://www.bloomberg.com/gadfly/articles/2018-04-19/advice-for-the-good-doctor-stop-smoking-cash.

Bloomberg. (2018b, June 15). Ping An Technology Stock Price. https://www.bloomberg.com/quote/1833:HK.

Bloomberg. (2018c, September 10). These are the economies with the most and least efficient health care. https://www.bloomberg.com/tosv2.html?vid=&uuid=b72af320-84e3-11eb-a334-d76e586ff4d2&url=L25ld3MvYXJ0aWNsZXMvMjAxOC0wOS0xLXMtbbmVhci1ib3R0b20tYtaGVhbHRoLWluZGV4LWhvbmecta29uZ-y1hbmQtc2luZ2Fwb3JlLWF0LXRvcA==.

Bloomberg News. (2022, August 9). Tiktok owner ByteDance pays $1.5 B for China Hospital chain in Health Foray. Bloomberg.com. https://www.bloomberg.com/news/articles/2022-08-09/bytedance-pays-1-5-billion-for-hospital-chain-in-health-foray#xj4y7vzkg.

Boulding, K. E. (1966). The concept of need for health services. The Milbank Memorial Fund Quarterly, 44(4): 202. https://doi.org/10.2307/3349064.

Bumgarner, J. M., Lambert, C. T., Hussein, A. A., Cantillon, D. J., Baranowski, B., Wolski, K., Lindsay, B. D., Wazni, O. M., and Tarakji, K. G. (2018). Smartwatch algorithm for automated detection of atrial fibrillation. Journal of the American College of Cardiology, 71(21): 2381–2388. https://doi.org/10.1016/j.jacc.2018.03.003.

Busse, R., Blümel, M., and WHO. (2014). Germany: health system review. Health Systems in Transition. https://apps.who.int/iris/bitstream/handle/10665/130246/HiT-16-2-2014-eng.pdf.

China News. (2018, June 11). China News, Ping An Medical Insurance Cloud Computing Healthcare Platform Update and Expansion, in Chinese "平安医保科技 "卫健云" 全面升级上线. http://sh.chinanews.com/swzx/2018-06-11/40394.shtml.

China Security Regulatory Commission. (2018, June 4). Pitch Book for Xiaomi CDR based Stock Offering. http://www.csrc.gov.cn/pub/zjhpublic/G00306202/201806/P020180611106685793601.pdf.

Cicchetti, A., and Gasbarrini, A. (2016). The healthcare service in Italy: Regional variability, European Review for Medical and Pharmacological Sciences. European Review. https://www.europeanreview.org/wp/wp-content/uploads/The-healthcare-service-in-Italy-regional-variability.pdf.

Dameff, C., Clay, B., and Longhurst, C. A. (2019). Personal Health Records. JAMA, 321(4): 339. https://doi.org/10.1001/jama.2018.20434.

Deloitte. (2018). Global health care outlook the evolution of smart health car. https://www2.deloitte.com/content/dam/Deloitte/global/Documents/Life-Sciences-Health-Care/gx-lshc-hc-outlook-2018.pdf.

Digital Transformation. (2021, March 8). Huawei. https://www.huawei.eu/what-we-do/digital-transformation.

Economist Intelligence Unit. (2016). Value-based healthcare in Europe: Laying the foundation. The Economist Intelligence Uni. EIU. https://eiuperspectives.economist.com/healthcare/value-based-healthcare-europe-laying-foundation.

Ehteshami Bejnordi, B., Veta, M., Johannes van Diest, P., van Ginneken, B., Karssemeijer, N., Litjens, G., van der Laak, J. A. W. M.; the CAMELYON16 Consortium; Hermsen, M., Manson, Q. F., Balkenhol, M., Geessink, O., Stathonikos, N., van Dijk, M. C., Bult, P., Beca, F., Beck, A. H., Wang, D., Khosla, A., Gargeya, R., Irshad, H., Zhong, A., Dou, Q., Li, Q., Chen, H., Lin, H. J., Heng, P. A., Haß, C., Bruni, E., Wong, Q., Halici, U., Öner, M. Ü., Cetin-Atalay, R., Berseth, M., Khvatkov, V., Vylegzhanin, A., Kraus, O., Shaban, M., Rajpoot, N., Awan, R., Sirinukunwattana, K., Qaiser, T., Tsang, Y. W., Tellez, D., Annuscheit, J., Hufnagl, P., Valkonen, M., Kartasalo, K., Latonen, L., Ruusuvuori, P., Liimatainen, K., Albarqouni, S., Mungal, B., George, A., Demirci, S., Navab, N.,

Watanabe, S., Seno, S., Takenaka, Y., Matsuda, H., Ahmady Phoulady, H., Kovalev, V., Kalinovsky, A., Liauchuk, V., Bueno, G., Fernandez-Carrobles, M.M., Serrano, I., Deniz, O., Racoceanu, D., and Venâncio, R. (2017). Diagnostic assessment of deep learning algorithms for detection of lymph node metastases in women with breast cancer. JAMA. 2017 Dec 12; 318(22): 2199–2210. doi: 10.1001/jama.2017.14585. PMID: 29234806; PMCID: PMC5820737.

EIU. (2016). China's healthcare challenges: The People's Hospital of Yiyang County in Henan Province, A case study from the Economist Intelligence Unit. http://vbhcglobalassessment.eiu.com/wp-content/uploads/sites/27/2016/10/EIU_Medtronic_CaseStudy_Chinas_Healthcare_Challenges.

Enthoven, A. C. (1978). Consumer-choice health plan. New England Journal of Medicine, 298(13): 709–720. https://doi.org/10.1056/nejm197803302981304.

European Commission. (2016, October). Health Care & Long-Term Care Systems, Germany, An excerpt from the Joint Report on Health Care and Long-Term Care Systems & Fiscal Sustainability. https://ec.europa.eu/info/publications/joint-report-health-care-and-long-term-care-systems-and-fiscal-sustainability-country-documents-2019-update_en.

European Commission. (2017). State of Health in the EU Country Health Profile 2017 Italy. https://ec.europa.eu/health/state/country_profiles_en.

European Commission. (2019a). Digital Single Market—Exchange of Electronic Health Records. https://ec.europa.eu/digital-single-market/en/exchange-electronic-health-records-across-eu.

European Commission. (2019b, June). eHealth strategies. country brief: Italy. https://ec.europa.eu/health/sites/health/files/ehealth/docs/ev_20190611_co922_en.pdf.

Eurostat. (2021). Healthcare expenditure statistics. https://ec.europa.eu/eurostat/statistics-explained/index.php?title=Healthcare_expenditure_statistics#Healthcare_expenditure.

Eversense. (2019). Long-term CGM System | Eversense Continuous Glucose Monitoring. https://www.eversensediabetes.com/eversense-cgm-system.

Fattah, S., Sung, N.-M., Ahn, I.-Y., Ryu, M., and Yun, J. (2017). Building IoT services for aging in place using standard-based IoT platforms and heterogeneous IoT products. Sensors, 17(10): 2311. https://doi.org/10.3390/s17102311.

Feldman, R., and Dowd, B. (1993). What does the demand curve for medical care measure? Journal of Health Economics, 12(2): 193–200. https://doi.org/10.1016/0167-6296(93)90028-d.

Ferré, F., de Belvis, A. G., Valerio, L., Longhi, S., Lazzari, A., Fattore, G., Ricciardi, W., and Maresso, A. (2014). Health system review. Health Systems in Transition: Vol. 16(4) [E-book]. WHO. https://www.euro.who.int/__data/assets/pdf_file/0003/263253/HiT-Italy.pdf?ua=1.

Financial Times. (2016). Ali Health: couples therapy. Retrieved from https://www.ft.com/content/4db3b2e8-d984-11e5-a72f-1e7744c66818.

Fitbit. (2019). Fitbit Comparison | Compare Fitness Trackers & Smartwatches. Retrieved April 10, 2019, from https://www.fitbit.com/global/us/compare.

Foundation, China Development Research. (2019). Reforming China's Healthcare System (Routledge Studies on the Chinese Economy) (1st ed.). Routledge.

Garmin. (2019). Fitness Watches | Sport Watches | Smartwatches | Garmin. Retrieved April 10, 2019, from https://buy.garmin.com/en-US/US/wearabletech/wearables/c10001-c10002-p1.html.

German Trade and Invest. (2017, February 15). The Mobile Health Market in Germany—Fact Sheets. GTAI. https://www.gtai.de/GTAI/Navigation/EN/Invest/Industries/Life-sciences/digital-health.html.

GKV. (2019, July). Responsibility for healthcare—The national association of the health and long-term care insurance funds. https://www.gkv-spitzenverband.de/media/

dokumente/presse/publikationen/GKV-Spitzenverband_Imagebroschuere_englisch_07-2019_barrierefrei.pdf.

Helissey, C., Parnot, C., Rivière, C., Duverger, C., Schernberg, A., Becherirat, S., Picchi, H., Le Roy, A., Vuagnat, P., Pristavu, R., Vanquaethem, H., and Brureau, L. (2023). Effectiveness of electronic patient reporting outcomes, by a digital telemonitoring platform, for prostate cancer care: the Protecty study. Front Digit Health. 2023 May 8; 5: 1104700. doi: 10.3389/fdgth.2023.1104700. PMID: 37228301; PMCID: PMC10203955.

Huawei. (2018). Healthcare—Huawei solutions. Huawei Enterprise. Retrieved June 25, 2019, from https://e.huawei.com/uk/solutions/industries/healthcare.

HUAWEI. (2018). Huawei Consumer Wearables. Retrieved June 25, 2018, from https://consumer.huawei.com/en/?type=wearables.

HUAWEI Accessories—HUAWEI Global. (2018). Huawei. Retrieved January 17, 2019, from https://consumer.huawei.com/en/accessories/.

Ip, J. E. (2019). Wearable devices for cardiac rhythm diagnosis and management. JAMA, 321(4): 337. https://doi.org/10.1001/jama.2018.20437.

ISTAT. (2017). Condizioni di Salute e ricorso ai servizi sanitari in Italia e nell'Unione Europea—Indagine EHIS 2015 [Health conditions and use of health services in Italy and the European Union—EHIS 2015 survey]. Rome: National Institute of Statistics (https:// www.istat.it/it/archivio/204655, accessed 2 November 2022).

Jakubowski, E., and Busse, R. (2002, May 31). Health care systems in the EU a comparative study : working paper. Publication Office of the European Union. https://op.europa.eu/en/publication-detail/-/publication/6125eac3-c1d3-4ea3-8f3d-9d9888d25e56.

Jing Dong Online Healthcare. (2019). Jing Dong Internet based Hospital. JD. Retrieved January 20, 2019, from https://care.jd.com/nh_home.

Jourdan, A. (2018, May 9). Tencent's WeDoctor raises $500 million, values firm at $5.5 billion pre-IPO. Reuters. https://www.reuters.com/article/us-china-tencent-wedoctor/tencents-wedoctor-raises-500-million-values-firm-at-5-5-billion-pre-ipo-idUSKBN1IA08G.

Kaminsky, L. (2019, January 17). The invisible warning signs that predict your future health. BBC Future. https://www.bbc.com/future/article/20190116-the-invisible-warning-signs-that-predict-your-future-health.

Kierkegaard, P. (2013). eHealth in Denmark: A case study. Journal of Medical Systems, 37(6): 1–10. https://doi.org/10.1007/s10916-013-9991-y.

Kringos, D., Boerma, W., Bourgueil, Y., Cartier, T., Dedeu, T., Hasvold, T., Hutchinson, A., Lember, M., Oleszczyk, M., Pavlic, D. R., Svab, I., Tedeschi, P., Wilm, S., Wilson, A., Windak, A., Van der Zee, J., and Groenewegen, P. (2013). The strength of primary care in Europe: an international comparative study. British Journal of General Practice, 63(616), e742–e750. https://doi.org/10.3399/bjgp13x674422.

Kyng, M., Manager, P., and Koester, T. (2014) Denmark—A Pioneer in Telemedicine Telemedicine Ecosystem.

Lovell, T. (2019). Germany introduces Digital Supply Act to digitalize healthcare. Retrieved July 14th 2023, from https://www.healthcareitnews.com/news/emea/germany-introduces-digital-supply-act-digitalise-healthcare.

Lee, E. (2020, September 28). JD Health preparing for Hong Kong listing · Technode. TechNode. https://technode.com/2020/09/28/jd-health-preparing-for-hong-kong-listing/.

Lepore, D., Dolui, K., Tomashchuk, O., Shim, H., Puri, C., Li, Y., Chen, N., and Spigarelli, F. (2023). Interdisciplinary research unlocking innovative solutions in healthcare. Technovation, 120: 102511. https://doi.org/10.1016/j.technovation.2022.102511.

Leichsenring, K. (2004). Providing integrated health and social care for older persons—A European overview. Providing integrated health and social care for older persons—A European overview of issues at stak. Aldershot: Ashgate Publishing Limited.

Leng, S. (2019, June 11). China's fragmented health care system under increasing pressure as nation rapidly ages. South China Morning Post. https://www.scmp.com/economy/china-economy/article/3013976/chinas-fragmented-health-care-system-under-increasing.

Lew, L. (2020, June 7). How Tencent's medical ecosystem is shaping the future of China's healthcare. TechNode. https://technode.com/2018/02/11/tencent-medical-ecosystem/.

Lim, J. (2014, June 17). WeChat is Being Trialled to make Hospitals more Efficient in China. Forbes. https://www.forbes.com/sites/jlim/2014/06/16/wechat-is-being-trialed-to-make-hospitals-more-efficient-in-china/#1db5084255e2.

Lo Scalzo, A., Donatin, A., Orzella, L., Cicchetti, A., Profili, S., and Maresso, A. (2009). Italy: Health system review. Health Systems in Transition: Vol. 11(6). WHO. https://www.euro.who.int/__data/assets/pdf_file/0006/87225/E93666.pdf.

Lovell, T. (2019, February 27). German health minister sets digitalisation as a top priority for new. Healthcare IT News. https://www.healthcareitnews.com/news/emea/german-health-minister-sets-digitalisation-top-priority-new-initiative?fbclid=IwAR34AD4DbmMh-3V9Z_9N-O8dnVgkRSbRG6DayvEs-CFTx5_rTxJfKjoa_0Y.

Mckinsey. (2010). China's Healthcare Reform. https://www.mckinsey.com/~/media/mckinsey/dotcom/client_service/healthcare%20systems%20and%20services/health%20international/hi10_china_healthcare_reform.ashx.

McKinsey & Company. (2019, December 2). How technology can improve the patient experience: A View from Tencent's Alex Ng. McKinsey & Company. https://www.mckinsey.com/industries/healthcare/our-insights/how-technology-can-improve-the-patient-experience-a-view-from-tencents-alex-ng.

Ministry of Health Denmark. (2017). Healthcare in Denmark, AN Overview. https://www.healthcaredenmark.dk/media/1479380/Healthcare-english-V16-decashx-3.pdf.

Ministero dell'Economia e delle Finanze. (2020). Ragioneria Generale dello Stato. Utilizzo dei Fondi strutturali e d'investimento europei. Rome: Ministry of Economy and Finance (https://senato.it/application/xmanager/projects/leg18/attachments/documento_evento_procedura_commissione/files/000/154/301/Contributo_dr._Castaldi.pdf, accessed 2 November 2022).

Ministry of Healthcare Italy. (2011). 7.10. Sistema di valutazione e monitoraggio della qualità dell'assistenza e delle performance dei sistemi sanitari - Relazione sullo Stato Sanitario del Paese. Ministry of Health Italy. http://www.rssp.salute.gov.it/rssp/paginaParagrafoRssp.jsp?sezione=risposte&capitolo=valutazione&id=2677.

MIT. (2019, May 7). Using AI to predict breast cancer and personalize care. MIT News|Massachusetts Institute of Technology. https://news.mit.edu/2019/using-ai-predict-breast-cancer-and-personalize-care-0507.

Murgia, M. (2019, July 31). DeepMind creates algorithm to predict kidney damage in advance. Financial Times. https://www.ft.com/content/c7cbcfba-b2b2-11e9-8cb2-799a3a8cf37b.

National Health Commission. (2021). Report of National Nutrition and Chronic Disease Status in China. http://www.nhc.gov.cn/mohwsbwstjxxzx/tjtjty/202106/984500aa83ac450e85c491a14aea3528/files/92e6c00f7dda4e9487303a09e942c06f.pdf

National Health Commission. (2018, June). Health System Review 2017 (in Chinese). http://www.nhc.gov.cn/guihuaxxs/s10743/201806/44e3cdfe11fa4c7f928c879d435b6a18.shtml.

National Health Commission. (2019, September 4). China starts mapping standards for 5G-based hospitals. http://www.gov.cn/guowuyuan/2019-09/04/content_5427271.htm.

National Health Commission. (2022, July). Health System Review 2021 (in Chinese). http://www.gov.cn/xinwen/2022-07/12/content_5700670.htm.

National Health and Family Planning Commission of the People's Republic of China. (2015, June). Report of National Nutrition and Chronic Disease Status in China. http://en.nhc.gov.cn/2015-06/15/c_45788.htm.

Obermann, K. (2013). Understanding the German Health Care System. Mannheim Institute of Public Health. http://www.goinginternational.eu/newsletter/2013/nl_03/SpecialDE_EN_Understanding_the_German.pdf.

OECD/European Observatory on Health Systems and Policies. (2017). Germany: Country Health Profile 2017. State of Health in the EU, 1–20. https://doi.org/10.1787/9789264283398-en.

Pacific Ventures. (2016, January 11). Electronic Health Records in China Provide Opportunities for International Healthcare Ventures. http://www.pcg-group.com/.

Pan, X.-F., Xu, J., and Meng, Q. (2016). Integrating social health insurance systems in China. The Lancet, 387(10025): 1274–1275. https://doi.org/10.1016/s0140-6736(16)30021-6.

Philips. (2019, October 18). The four pillars of a value-based healthcare future. https://www.philips.com/a-w/about/news/archive/future-health-index/articles/20180126-pillars-value-based-healthcare.html.

Ping An Good Doctor. (2019). Ping An Good Doctor. Retrieved September 16, 2019, from https://www.jk.cn/.

Rahman, M. M., Khatun, F., Uzzaman, A., Sami, S. I., Bhuiyan, M. A.-A., and Kiong, T. S. (2021). A comprehensive study of artificial intelligence and machine learning approaches in confronting the coronavirus (COVID-19) pandemic. International Journal of Health Services, 51(4): 446–461. doi:10.1177/00207314211017469.

Ram, A. (2018, April 5). Babylon signs Tencent deal to deploy health technology on WeChat. Financial Times. https://www.ft.com/content/40fae194-381d-11e8-8eee-e06bde01c544.

Rapid Value Solutions. (2020, November 3). Is Wearable Technology the Future of Healthcare? RapidValue. https://www.rapidvaluesolutions.com/wearable-technology-the-future-of-healthcare/.

Renew Health Project. (2009). Renew Health-Project Objective. https://www.consorzioarsenal.it/en/web/guest/progetti/renewing-health/gli-obiettivi.

Ridic, G., Gleason, S., and Ridic, O. (2012). Comparisons of Health Care Systems in the United States, Germany and Canada. Materia Socio Medica, 24(2): 112. https://doi.org/10.5455/msm.2012.24.112-120.

Rinbeat. (2018). Rinbeat – Intelligent Monitoring. http://rinbeat.it/.

Shen, J. (2020, June 8). Tencent doubles down on healthcare, tests medical services on WeChat · Technode. TechNode. https://technode.com/2019/04/08/tencent-healthcare-service-wechat/.

Schnall, R., Rojas, M., Bakken, S., Brown, W., Carballo-Dieguez, A., Carry, M., Gelaude, D., Mosley, J. P., and Travers, J. (2016). A user-centered model for designing consumer mobile health (mHealth) applications (apps). Journal of Biomedical Informatics, 60: 243–251. https://doi.org/10.1016/j.jbi.2016.02.002.

Sohu Finance. (2019). Ping An Good Doctor service reaching high way service stations. https://www.sohu.com/a/292263887_115865?scm=1002.44003c.fd00fe.PC_ARTICLE_REC.

Stanford Medicine. (2018). The Democratization of Health Care', Standford Medicine 2018 Health Trends Reports. https://med.stanford.edu/content/dam/sm/school/documents/Health-Trends-Report/Stanford-Medicine-Health-Trends-Report-2018.pdf.

State Council. (2016, January). Opinions on Integrating medical insurance schemes for rural and urban residents, in Chinese 国务院关于整合城乡居民基本医疗保险制度的意见_政府信息公开专栏. http://www.gov.cn/zhengce/content/2016-01/12/content_10582.htm.

State Council. (2018, April). Opinions on Promoting Internet + Healthcare Industry development, in Chinese 国务院办公厅关于促进"互联网+医疗健康"发展的意见（国办发(2018)26号）_政府信息公开专栏. http://www.gov.cn/zhengce/content/2018-04/28/content_5286645.htm.

Stern, A. D., Brönneke, J., Debatin, J. F., Hagen, J., Matthies, H., Patel, S., Clay, I., Eskofier, B., Herr, A., Hoeller, K., Jaksa, A., Kramer, D. B., Kyhlstedt, M., Lofgren, K. T.,

Mahendraratnam, N., Muehlan, H., Reif, S., Riedemann, L., and Goldsack, J. C. (2022). Advancing Digital Health Applications: Priorities for innovation in real-world evidence generation. The Lancet Digital Health, 4(3). https://doi.org/10.1016/s2589-7500(21)00292-2.

Subramania, J. (2019, March 25). Why IoT is not a technology solution—it's a business play. Microsoft Azure. https://azure.microsoft.com/en-us/blog/why-iot-is-not-a-technology-solution-it-s-a-business-play/.

Sun, J., Lin, Q., Zhao, P., Zhang, Q., Xu, K., Chen, H., Hu, C. J., Stuntz, M., Li, H., and Liu, Y. (2017). Reducing waiting time and raising outpatient satisfaction in a Chinese public tertiary general hospital-an interrupted time series study. BMC Public Health, 17(1): 9–21. https://doi.org/10.1186/s12889-017-4667-z.

Tang, P. C., Ash, J. S., Bates, D. W., Overhage, J. M., and Sands, D. Z. (2006). Personal health records: Definitions, benefits, and strategies for overcoming barriers to adoption. Journal of the American Medical Informatics Association, 13(2): 121–126. https://doi.org/10.1197/jamia.m2025.

Tencent Doctorwork. (2018). Tencent Doctorwork. https://www.doctorwork.com/index/pc/.

Terry, N. (2016). Will the Internet of Health Things Disrupt Healthcare? SSRN Electronic Journal, 327–352. https://doi.org/10.2139/ssrn.2760447.

The Economist. (2017, May 11). China needs many more primary-care doctors. https://www.economist.com/china/2017/05/11/china-needs-many-more-primary-care-doctors.

The Economist Intelligence Unit. (2015). Value-based health assessment in Italy A decentralised model. EIU. https://eiuperspectives.economist.com/healthcare/value-based-health-assessment-italy-decentralised-model.

The German Federal Institute for Drugs and Medical Devices. (n.d.). BfArM - About us. BfArM. Retrieved September 5, 2019, from https://www.bfarm.de/EN/BfArM/_node.html.

The System of Health accounts in Italy. (2017, July 4). ISTAT. https://www.istat.it/en/archive/201949.

Wang, S., Fu, X., Liu, Z., Wang, B., Tang, Y., Feng, H., and Wang, J. (2018a). General Practitioner education reform in China: Most undergraduate medical students do not choose general practitioner as a career under the 5+3 Model. Health Professions Education, 4(2): 127–132. https://doi.org/10.1016/j.hpe.2017.05.001.

Wang, Z., Chen, Z., Zhang, L., Wang, X., Hao, G., Zhang, Z., Shao, L., Tian, Y., Dong, Y., Zheng, C., Wang, J., Zhu, M., Weintraub, W. S., and Gao, R. (2018b). Status of Hypertension in China. Circulation, 137(22): 2344–2356. https://doi.org/10.1161/circulationaha.117.032380.

Wendt, C. (2009). Mapping European healthcare systems: a comparative analysis of financing, service provision and access to healthcare. Journal of European Social Policy, 19(5): 432–445. https://doi.org/10.1177/0958928709344247.

WHO. (2003). The Domains of Health Responsiveness A Human Rights Analysis. https://www.who.int/healthinfo/paper53.pdf.

World Health Organization. Regional Office for the Western Pacific. (2015). People's Republic of China health system review. Health Systems in Transition, 5(7). WHO Regional Office for the Western Pacific. https://apps.who.int/iris/handle/10665/208229.

WHO. (2018, June). Towards a roadmap for the digitalization of national health systems in Europe. World Health Organization. https://www.euro.who.int/__data/assets/pdf_file/0008/380897/DoHS-meeting-report-eng.pdf?ua=1.

Wong, J. C. (2019, July 17). Elon Musk unveils plan to build mind-reading implants: "The monkey is out of the bag." The Guardian. https://www.theguardian.com/technology/2019/jul/17/elon-musk-neuralink-brain-implants-mind-reading-artificial-intelligence.

World Bank. (2022). Age dependency ratio (% of working-age population) - China | Data. World Bank Data Bank. https://data.worldbank.org/indicator/SP.POP. DPND?locations=CN.

World Bank. (2022). Out-of-pocket health expenditure as a percentage of current health expenditure (%). Retrieved from https://data.worldbank.org/indicator/SH.XPD.OOPC. CH.ZS.

World Bank. (2019). Urban population (% of total population) - China. Retrieved [Feb 2nd 2019], from https://data.worldbank.org/indicator/SP.URB.TOTL.IN.ZS?locations=CN.

World Economic Forum & Boston Consulting Group. (2017). Value in Healthcare Transformation for Health System Laying the Foundation. World Economic Forum. http://www3.weforum.org/docs/WEF_Insight_Report_Value_Healthcare_Laying_ Foundation.pdf.

Wuzhen Internet Hospital. (2019). Wuzhen Internet based Hospital, 乌镇互联网医院B端业务，已赋能1200多家县域医院，构建基层医疗新生态-动脉网. VC Beat. https://vcbeat. top/ODAxNDhjNDU0YmFkZGViYzU5MmQ5ODNmYTM4ZThmZDY=.

Xiaomi. (2018, June 15). Xiaomi Market. https://www.mi.com/index.html.

Xiaomi. (2019). Mi Band 4 (in Chinese). Xiaomi Mi Band 4. https://www.mi.com/ shouhuan4nfc.

Zhou, M., Wang, H., Zeng, X., Yin, P., Zhu, J., Chen, W., Li, X., Wang, L., Liu, Y., Liu, J., Zhang, M., Qi, J., Yu, S., Afshin, A., Gakidou, E., Glenn, S., Krish, V. S., Miller-Petrie, M. K., Mountjoy-Venning, W. C., Mullany, E. C., Redford, S. B., Liu, H., Naghavi, M., Hay, S. I., Wang, L., Murray, C. J. L., and Liang, X. (2019). Mortality, morbidity, and risk factors in China and its provinces, 1990–2017: a systematic analysis for the Global Burden of Disease Study 2017. Lancet. 2019 Sep 28; 394(10204): 1145–1158. doi: 10.1016/ S0140-6736(19)30427-1. Epub 2019 Jun 24. Erratum in: Lancet. 2020 Jul 4;396(10243):26. PMID: 31248666; PMCID: PMC6891889.

华为穿戴 – 华为官网. (2019). Huawei Wearables. Retrieved April 10, 2019, from https:// consumer.huawei.com/cn/wearables/.

CHAPTER 2

Developing the Innovation Ecosystem

A Case Study on the Chinese Digital Healthcare Industry

1. Introduction

The theoretical framework supporting the digital ecosystem development of emerging markets needs an update to correspond with the new empirical evidence in business models and governance structures used by tech firms from emerging markets such as China. Chinese firms were chosen as the target of the case study on the smart health industry as that market is highly capitalized and competitive (Financial Times 2020). Novel and flexible business models and financial schemes have thus been created to enable firms to tackle intense competition and engage customers.

Not so long ago, the smartphone market was still dominated by Apple, Samsung, Nokia and Blackberry; Chinese companies such as Huawei or Xiaomi were unknown to customers. Today, these two companies are the world's leading smartphone manufacturers (IDC Market Research 2020). Other Chinese companies, such as Alibaba, were once seen as the copycats of Ebay and Amazon. The Alibaba ecosystem is now going beyond e-commerce where the business originated from; it has created and embedded a business model and ecosystem envied by its western counterparts. Nowadays, Alibaba is leading in the entertainment and finance sectors in China; meanwhile, it has started venturing into healthcare, an area with sophisticated technologies and added value, with AI and cloud computing.

For the Internet of Healthcare Things[1] industry, Xiaomi and Huawei are among the world's top five wearable producers by shipment volume (IDC 2019). Both of these companies made heavy investments in sports and healthy living lifestyle-related products. Alibaba has developed solutions for both individual users and institutional healthcare service providers. These solutions focus on reducing waiting time for healthcare services, improving healthcare quality and promoting healthcare efficiency. More specifically, their work focuses intensely on lowering the processing time of medical images and improving diagnosis efficiency and accuracy for healthcare organizations; for individual customers, their services largely consist of selling healthcare services and products online. The Alibaba online-to-offline medicine sale is growing rapidly, with sales from online-to-offline medication jumping by 296.8% from 2017 to 2018 (Alibaba Health Information Technology Limited 2019). Private insurance companies such as Ping An, and public insurance schemes are now barging in to acquire and analyze data on a population health management scale, and offer policy advisory on public healthcare issues. The outbreak of coronavirus has proven the essentiality and boosted the demand for smart healthcare services.

This chapter aims to elaborate on how Chinese firms have transferred their competitive advantage to new sectors such as healthcare and AI, based on the digital ecosystem platform theory and the emerging market internationalization theory. A theoretical framework based on the case studies from Alibaba, Huawei and Xiaomi has been established herein. These case studies have been developed from interviews conducted with investment and business managers in leading Chinese tech firms such as Alibaba, Huawei and Xiaomi.

2. Literature Review

2.1 Digital platform ecosystem theory

Using Google Scholar and searching for past literature on the subject of digital platform ecosystems, comparative advantage and emerging market enterprises, literature was surveyed. Literature published in 2003 was referenced as it was a highly downloaded and referenced article.

2.1.1 The dynamic capabilities of Digital platform ecosystems

Digital platform ecosystems (DPEs) are leading the technological revolution in our era (Teece 2018a,b). DPEs are often multifaceted providing

---

[1] Internet of Healthcare Things (IoHT) refers to identifiable devices connected to the Internet and able to communicate with each other, used for healthcare purposes. Comarch 2020, available at: https://www.comarch.com/iot-ecosystem/internet-of-healthcare-things/.

interfaces for two or more groups of economic actors from different sides of the platform (Healfat and Raubitschek 2018). The DPEs refer to the ones which facilitate and mediate transactions and communications between actors. A multifunctional digital ecosystem consists of a platform leader, actors on different sides of the platform, and infrastructure providers for the platform (Healfat and Raubitschek 2018).

Healfat and Raubischek (2018) pointed out that multi-sided digital ecosystems can generate cross-side effects for users. The service quality of the platform and hence the value for one side of parties depends on the number and quality of parties on the other side of the platform. Cross side effects of digital platforms can be both positive and negative. Healfat and Raubischek (2018) identified three types of dynamic characteristics critical for platform leaders: innovation ability, environmental scanning and sensing ability, and integrative ability for ecosystem organization.

Based on dynamic capabilities according to Teece (2007), Healfat and Raubischek (2018) defined three dynamic characteristics necessary in generating strategic changes for DPEs: (a) identifying new market opportunities and threats, (b) making use of new opportunities with innovation in business models and strategic investments, and (c) adapting the existing business models and strategies.

According to Helfat and Raubitschek (2018), the three capabilities are essential for profiting from innovation in digital platform-based ecosystems. The authors also distinguished between general capabilities and dynamic capabilities. Whereas general capabilities refer to the ones related to regular operations of business, the latter differs from them in that dynamic capabilities lead to strategic changes at organizational as well as individual levels (Helfat and Raubischek 2018). The authors pointed out that dynamic capabilities are the ones which allow firms to create, extend and modify the value creation process. These capabilities can enable changes internally and externally, e.g., modifications in the business model, intangible and tangible assets and operational capabilities. This differentiates dynamic capabilities from the capabilities to operate the business on a daily basis (Helfat and Raubitschek 2018).

(1) Innovation capabilities

By innovation, Healfat and Raubischek (2018) referred to product sequencing, which entails linking new and existing product, services, associated knowledge and capabilities in time and over time. The product sequencing process does not include upgrading just the core products in the ecosystem, but also the features of the whole ecosystem. Multi-team software innovation capabilities of platform leaders are necessary for innovations in the ecosystem. Platform leader, which sits on the top of the ecosystem, can benefit from routines with cross-functional teams and the ability to coordinate between the innovation teams.

(2) Sensing/scanning capabilities

On the institutional level, sensing capabilities require the platform leader to know the core as well as the complementary products. The ability to gain market access (Helfat and Winter 2011), which include those for interacting and obtaining feedback from customers through sales and service teams, is a part of the environmental scanning function. On an individual level, top managers can also sense opportunities and threats from the feedback received from customers, stimulating the product sequencing process. The scanning and sensing capabilities are key for identifying market threats such as complementary asset providers for competitors, or for identifying market opportunities such as creating new channels to reach the customers.

(3) Integrative capabilities for resource orchestration

Besides creation, placing the new product and setting the platform rules, platform leaders are also engaged in coordinating resources around DPE (Helfat and Raubischek 2018). Strong digital ecosystems are characterized by cross-side network effects. For platform leaders, the success of the business model is heavily dependent on the number of users and the quality of contents on both sides of the platform. Network effects determine the value of the platform. By organizing the governance rules for ecosystem members, the DPEs ensure the quality of complimentary resources they provide. By routinely communicating with different teams, the platform leader can organize internal resources efficiently, detect problems and make the required strategic changes. In the process, DPEs may get involved in new asset acquisition, and assembling and coordination of resources for the platform. Figure 2.1 explains the important components of the digital platform ecosystem theory.

DPEs differ from traditional multinationals, the former being more adaptive to market needs. This characteristic makes it relatively easy for DPEs to access new markets, adapt to existing markets and get rid of products with low market shares and low returns. The product, or the value created, exists in a digital format, which can be easily tested with customers and changed accordingly. The time it takes for DPEs to adjust strategies, and the due diligence process take less time than traditional multinational enterprises (MNEs). This makes the business model of most DPEs more flexible than that of traditional MNEs It is also easier for DPEs to access a foreign market.

As DPEs, companies are able to be 'born global', with the product often launched in several markets and platforms at once. With a large customer base, it also becomes relatively easy to get new partners on board (Nambisan et al. 2019). According to Sharma and Blomstermo (2003), the born global multinationals follow an internationalization model which supports the

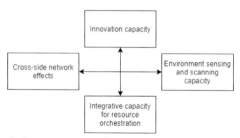

Figure 2.1: Digital platform ecosystem theory, author's illustration. Source: Teece 2017, Helfat and Raubischek 2019.

path dependency theory. To elaborate, the internationalization entry model and established internationalization networks are usually based on weak ties and knowledge that firms gather in the domestic market and the network established in the domestic market. The network refers to the number of users accumulated in the domestic market as a resource which can be leveraged for partners in the foreign market; knowledge refers to the capability to improve the attractiveness of the platform and get more users on board.

With not many tangible assets, DPEs do not need to be physically present in a market to access customers. With a large customer base, firms without previous internationalization experiences can also access international customers via DPEs. This is particularly beneficial for start-ups and small and medium size firms with limited resources. DPEs provide SMEs with a ready to function platform which allows SMEs to access the global customer resources (Nambisan et al. 2019). For manufacturing firms, it is also easier to find suppliers for services with DPE platforms.

2.1.2 Digital Platform Ecosystems and network effects

DPEs offer a ready to plug-in platform with a large customer base to MNEs, SMEs, start-ups and individual suppliers (Nambisan et al. 2019). With all the data on supply and demand, DPEs make it easier for firms to locate market demand. This challenges the traditional market order, such as the parent-subsidiary ranks, making the emergence of multiple cost centers independent from one another possible (Nambisan et al. 2019). Social networks such as Twitter, Facebook, Instagram, Weibo and Tik Tok offer brands more opportunities to create virtual online campaigns and interact with customers directly.

The vast network created by DPEs allows participants to adjust to market needs in a more flexible manner. As a collective actor, DPEs often have the operational and structural flexibility in dealing with market turbulences. This allows DPEs to identify new opportunities in the market, which would benefit individual members of the platform.

Banalieva and Dhanaraj (2019) tried to explain that digital networks have become one of the firm-specific critical advantages for digital service multinational firms. Networks have become a mode of governance as well as strategic resources for digital platforms. The advantages brought by networks are strategically important for DPEs, compared to the traditional asset-based systems. Banalieva and Dhanaraj (2019) made the following arguments regarding the role of digitalization in firms' internationalization process.

Firstly, digitalization enables firms to separate firm-specific advantages into technology and human capital. This makes it easier for them to reconcile between advanced skills and common and more generic skills in the foreign market.

Secondly, digitalization increases the transferability of the firm-specific advantages of a firm's technology—by enhancing the technology's modularity and promoting bundling with local firms. Module here refers to the integration of multiple layers of technology. As advanced skills such as coding for algorithms is hardly obtainable in certain developing markets, digital platforms make it easier to deploy algorithms in the foreign market, employing local employees with more generic skills (Banalieva and Dhanaraj 2019).

Thirdly, digital platforms may increase the risk for firm-specific advantages to be copied across borders. Meanwhile, digitalization also enhances a firm's ability to limit imitation by integrating complex technological proprietary and network effects.

Fourthly, digitalization increases a firm's ability to bundle its advanced (or generic) human capital firm-specific advantages. Digitalization may increase specification of advanced human capital skills and demote firm-specific generic human capital skills.

Fifth, digitalization enhances a firm's ability to exploit its core (or peripheral) technical firm-specific advantages in foreign markets with a more (or less) centrally controlled digital network.

Lastly, digitalization enhances a firm's ability to exploit advanced (or generic) human capital firm-specific advantages in foreign markets at lower costs.

Overall, networks, and hence the number of users and the quality of contents on the platform, have become a competitive advantage for multinational digital service firms. Large networks provide more diverse datasets which enable more precise business insights, compared to smaller networks with non-integrative datasets. To summarize, Banalieva and Dhanaraj (2019) claimed that networks connect organizational design, theorize asset-light internationalization, enable firms to make governance choice predictions, and increase knowledge transferability as well as the risk of appropriability. They tested their theory based on the emergence of network-based platforms.

Hennart (2019) pointed out that contrary to what Banalieva and Dhanaraj (2019) argued, networks are not governance structures. Hennart (2019) insisted on using the traditional international business theory such as the OLI paradigm (Ownership, Location, Internationalization) proposed by Dunning and Ludan (2008) to explain digital platforms going abroad. Hennart (2019) used the case of McDonald's franchising all chain restaurants to explain the relationship between UBER and its contracted drivers. In such a hierarchical relationship, Uber controls the behavior of the drivers by deploying controlling mechanisms in the app and discontinuing the contract with the driver if necessary.

According to Hennart (2019), the network exists as there are interdependencies between firms on the value chain; multinational digital platforms try to internalize the cost of transactions (the exchange of outputs) and thereby establish the network on the value chain. Digital platforms, like traditional firms, try to protect their reputation via trademarks. They still follow traditional business models such as B2C, B2B, C2C and C2B, while networks exist in the business model on the industry level. Besides, the effects of the network model can only be observed ex-post instead of ex-ante. Network effects can only work for new entrants to the sector when switching costs are high. Therefore, network effects cannot be calcified as a firm-specific advantage.

There are valid points made in both the studies. Two-sided or multi-sided network effects and Moore's Law are two phenomenon which determine the success of tech firms whether in Silicon Valley or in Shenzhen, China. However, whether network effects and Moore's Law can become firm-specific advantages, it depends on a lot of factors. Platforms such as Facebook or LinkedIn dominate the social network with over 2 billion monthly active users (MAU) and 303 million MAU respectively. The switching costs are already high enough that anyone who wishes to shed their privacy is immediately excluded from the network provided by these two platforms.

The three authors try to use examples from developed economies, such as Uber and McDonald's, to validate their theories. Digital platforms from emerging markets, such as Didi and Alibaba in Hennart (2019), were but briefly mentioned. However, there was a lack of deep empirical work based on emerging market digital platforms and comparing their business models with their counterparts in developed markets.

2.2 Emerging market multinationals (EMMs) internationalization theory

2.2.1 The defective path-and-context based framework

The traditional OLI framework developed by Dunning, which explained foreign direct investment (FDI), was developed during 1950s and 1970s

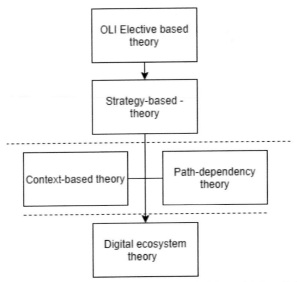

Figure 2.2: The evolution of International Business Models explaining digital platforms; Source: Dunning (2001, 2013), Deng et al. 2020, Knoerich 2019, Liu 2019, Nambisan et al. 2019, Hennart 2019.

(Knoerich 2019). These theories were formed to explain multinational firms from developed countries making investments in developing economies. The multinationals were mainly seeking low labor costs and large markets in developing economies. Today, foreign direct investments made from emerging market economies in the overseas market are obviously made for different reasons in very different settings.

Today, the dominating theories seem to be path-dependent theories where authors try to explain foreign direct investment firms in a contextual and institution-based manner (Liu 2019, Deng 2020). However, based on the summary of the literature listed below, the authors are still trying to understand the changing political, economic, social and technological settings in developing economies and their interactions with specific firms. Therefore, here, we try to contradict and compete with the existing theories regarding EMMs making OFDI decisions from the national, industrial and firm levels, explaining the changes in political, economic, social and technological settings in China and the impact they have on tech firms (from the semi-conductor, AI and internet industries) when they are going abroad.

As emerging markets digital platforms (EMDPs) grow stronger, they have started to navigate foreign markets, for domestic markets gradually become saturated, mature and competitive. EMDPs expand overseas for a larger customer base, stronger domestic brand perception, obtaining international talents, and adapting to and subsequently shaping international standards. The existing theories consider expanding overseas as a strategic decision and did not take into account the unique features of digital platform ecosystems in terms of their appetite for more users, content, algorithm and design. The conditions determining the development of digital ecosystems on institutional, industry and national levels are different from traditional businesses expanding into overseas markets.

Hereby, a summary of traditional theories explaining EMMs and their OFDI strategy has been provided. In the next part, the features of digital ecosystem platforms and their motivations for going overseas will be elaborated.

Deng et al. (2020) identified five characteristics of EMMs:

The first characteristic identified was a weak domestic institutional setting. The authors used an example of Huawei, explaining how the company could not separate itself from the state.

Secondly, the authors identified a strong government influence in the firms' decision to go abroad. The Belt and Road initiatives can be a case which prompts state-owned firms to explore the overseas market.

Thirdly, the lack of superior technological and managerial resources as one of the characteristics for EMMs to go abroad was identified by the authors.

Fourthly, the authors pointed out that MNCs were going through an early stage of internationalization. By early stage, the authors referred to a lack of superior technological and managerial resources as the key feature of EMMs.

Fifthly, the authors inferred close ties to the location of their origin as one of the key features of emerging market multinationals.

There are several valid points made by Deng et al. (2020), highlighting that the internationalization of EMMs contradicts the traditional theory of OLI paradigms. Indeed, the OLI paradigms were developed in the 1960s, based on the traditional internationalization path of multinationals that originated developed countries (Knoerich 2019). These multinationals typically sought investments for labor intensive and cost intensive markets which could lower the production cost. This has shaped the formation of the global value chain as it stands today. The context-based framework developed by Deng et al. (2020) partially makes sense, given the geopolitical environment that EMMs are in today, such as the impact of trade wars between U.S. and China. There are some points worth further discussion—such as weak domestic institutions and lack of superior technological and managerial resources as a characteristic of emerging market multinationals going abroad.

First of all, China, has shown strong institutional capacity in the domestic setting by supporting large-scale projects; some scholars might argue that the competitors of Chinese companies, such as Google, Amazon, Apple, Ericsson and Nokia, have stronger domestic institutional settings. However, when it comes to economy of scale, big Chinese tech companies enjoy the huge domestic market, supportive government policies such as tax subsidies, full supply chain, and a complete set of infrastructure (water, electricity, 5G network, credit lines, etc). Even foreign tech companies such as Tesla got the first car out of the production line from the manufacturing plant 'Mega Factory' in Shanghai in December 2019 (Financial Times 2019); It took 168 days for Tesla to get the permits for construction and get a finished factory. Compared to the fate of the Tesla factory in Berlin, the Chinese government has offered cheap financing from state-backed lenders, amounting to RMB 11.25 bn (USD 1.6 bn) for the Mega Factory in Shanghai(Hull and Zhang 2019). Strong institutional setting and supportive industrial policies are indeed necessary for economic development. In part 3, policy settings towards wearables industry development will be discussed in details.

Secondly, given the high level of maturity of the Chinese tech market and the intensive competition, Chinese companies which went overseas are seeking alternative markets and higher returns overseas. Most Chinese companies that have made foreign investments are survivors and have often acquired a dominant position in the Chinese domestic market. Usually, the global headquarters of Chinese multinationals are still based in China. The Chinese headquarter serves as the coordinator of global resources and as the engine for driving global innovations for different types of businesses as well. The overarching concept that emerging market multinationals do not have superior technological and managerial resources should hence be rectified.

Thirdly, high-profile technological companies such as Alibaba, Tencent, Huawei and Xiaomi are privately owned in China, as opposed to state-owned firms; the motivations of foreign direct investment by Chinese firms remain profit driven. In the process, Chinese companies may be associated with the Chinese government and its motivations. This leads to conflicts of state interests and business interests. Do government policies such as One Belt One Road (OBOR) play a role in Chinese foreign investment? To a certain degree, yes. However, OBOR projects are mostly related to infrastructure. Tech companies rarely make investments in regions such as central Asia, Pakistan or Horn of Africa to align themselves with state interests. Instead, all the investments made by tech companies are related to strategic development and ecosystem development.

Fourthly, the success of Chinese firms going global started with Haier, a privately owned home appliances business Haier started investing in the U.S. in greenfield style, with an industrial park in 1999 (Haier 2019). 20 years have passed and Haier has made M&A in New Zealand and Japan and its products are available in China, U.S. and Europe. In the age of IoT, Haier has also shown off its pilot connected kitchen in CES Asia 2019. Haier has won the award for innovation in management theory and framework at the National Management Conference as well. Haier, as a Chinese company, has accumulated 20 years of knowledge and network in internationalization. Lack of superior technological and managerial resources does not seem to fit in the case of Haier.

Fifthly, any multinational going abroad maintains close ties with their origin. Take tech companies from the U.S. for instance; Google, Amazon and Facebook were recently heavily fined by the European Commission for breaking the anti-competition, tax and privacy related laws and regulations. The then U.S. president, Donald Trump, not only showed his support by sending several tech officials to CES, but also blamed the European Commissioner, Margrethe Vestager, saying that "She hates the U.S." (Financial Times 2020). By supporting Qualcomm in the court case for patent, and sending Huawei to court for stealing 'U.S. Technology' (Financial Times 2020), the U.S. government certainly supports U.S. companies in building 5G networks around the globe by bending the U.S. legal system towards state interests.

2.2.2 Buckley and the three layers of factors affecting firm internationalization

Buckley (2019) identified three layers of factors affecting Chinese firms going abroad. The author considered national (provincial and city) level policy factors, industrial level factors and firm level factors. The author identified the historical phases of Chinese firms going abroad, with phase one starting in 1979, with the Chinese government's opening door policy, phase two involving government encouragement, phase three concerned with expansion and regulation, phase four being the go global phase, and phase five—the post-WTO stage. The author identified the trends in overseas foreign direct investment in the format of M&A, as made by Chinese firms.

Firstly, Buckley (2019) seems to consider the defected Chinese capital market as one of the factors affecting the overseas M&A activities of Chinese firms. The abundance of Chinese capital invested in advanced economies such as the UK, Australia, Canada and the U.S.A. is attracting local media and government attention towards the Chinese seeking strategic assets or stealing foreign technology. This on a national level causes counter-Chinese investment measures in the U.S., Australia and the UK (Financial Times 2019)—on using Chinese suppliers for 5G network construction. On an industrial level, Huawei was nearly banned as an IEEE reviewer and from participating in standard setting on 5G. On a company level, the Huawei CFO was arrested in Canada, which on a personal level attacked the founder of Huawei—the father of the CFO. This proves that the interests of Chinese private businesses and Chinese national interests do not align. As a communist country, China is seen as a rival rather than an ally in most advanced economies. With the rise of populism, in China, investments are viewed as threats rather welcomed boost to the local economy.

Secondly, Buckley (2019) suggests that Chinese firms make overseas investments in markets with close ties to the Chinese government, with high risks. Considering the large-scale infrastructure projects in Myanmar, Pakistan and Ethiopia, this may be true. When it comes to the foreign direct investments made by tech firms, the incentive for COFDI is always related to profit. One of the biggest incentives is to acquire new markets, given that the domestic Chinese market is highly mature with abundant tech infrastructures in urban regions and has intensive competition. Xiaomi is marketing in the rural areas of China. Successful tech companies, such as One Plus, chose to expand in South East Asian markets like India. Huawei set up R&D centers in Europe to acquire and use foreign design talents; interestingly, most foreigners view Huawei as a threat, yet the infrastructure of T Mobile, Vodafone and several other telecommunication providers are all equipped with Huawei switches. Huawei went to the African market to obtain market shares and dominant positions, and become market leaders. The other purpose was to train the talents of Huawei in one of the most difficult environments on the earth to do business—Uganda (Parkinson et al. 2019, World Bank Doing Business Index 2020). Uganda is a low-income country in Africa, ranking 169th on the World Bank Doing Business Index. In contrast with what western media described as providing tools for surveillance of its citizens (Parkinson et al. 2019), Huawei provides critical communications infrastructure to the population in Uganda. Without companies such as Huawei, the African population cannot use basic mobile services.

Thirdly, Buckley (2019) noticed that COFDI (Chinese outward foreign direct investment) flows to tax havens. The situation is explained by the ineffectiveness of the Chinese capital market and the attenuation effects of tax havens in orchestrating resources between different locations. When it comes to the three surveyed firms—Huawei, Xiaomi and Ali-health, Huawei is registered in Shenzhen, China, while Xiaomi is registered in the Cayman Islands and Alihealth is registered in Bermuda. Barring Huawei, both Xiaomi and Alihealth are registered in tax havens. This may relate to the fact that Huawei is a privately owned firm, while Alihealth and Xiaomi are publicly listed in Hong Kong. Strategically, public listed companies need to display their revenues and earnings in their quarterly report. This may force these firms to focus more on short-term interests if they do not have enough cash holdings to counteract the fluctuations in the stock market.

Fourthly, Buckley (2019) argues the lack of transparency as one of the key features of Chinese firms' overseas M&A. In case of state-owned and privately owned firms, this might be true. However, when it comes to publicly listed firms, all their behaviors must be reported to the stakeholders quarterly. Therefore, this assumption does not apply to most tech firms in China—listed overseas or domestically.

2.2.3 The benefits of overseas M&A: exchange of knowledge and networks

He et al. (2019) identified technology as one of the firm-level advantages for the internationalization of Chinese firms. He et al. (2019) used three case studies to confirm his theory, namely BYD, Sany Heavy Industry and CSR China. All of these three are heavy manufacturing firms in automobile (electric cars/buses/railway cars and electronics), construction equipment and railway/metro cars industry. He et al. (2019) used secondary data to support his theory of Chinese firms' internationalization. Among them, BYD, a privately owned firm, started as a battery manufacturer and expanded business into electric cars. Now, it has expanded from China to the Netherlands, Singapore, U.S.A., Brazil, Ecuador, Chile and Colombia. Sany, again a privately-owned Chinese firm, expanded to Russia, Latin America, South America and Africa. CSR China, a state-owned firm, now has production lines in the U.S. and has expanded business there as well. All the data collected was second handed in the He et al. (2019) paper. Undeniably, a few years ago, Chinese metro and railway construction was still using technology from Hong Kong and Germany. Now Chinese firms are exporting such technology to Africa, the Netherlands and U.S.A, proving that Chinese firms can learn, transform and provide low-cost solutions fast.

An interesting fact here is that the western media cannot separate private Chinese firms and state-owned Chinese firms. They believe that all firms in China are related to the Chinese government and the Chinese government has control of all data collected by all the firms in China, according to the national security law of the Chinese government. In reality, Chinese firms go abroad with the same incentives as their western counterparts—to acquire users, expand their product ecosystem and acquire knowledge and networks.

Liu (2019) used the M&A case of Goldwind in Germany and Envision Energy Co., Ltd. as examples of Chinese firms catching up in the manufacturing of wind turbines. Both of these case studies were conducted based on interviews with managers from Goldwind and

Envision. The only difference was that Goldwind was an established firm, while Envision started with a few investment managers from London, with the ambition and vision to build wind turbines; the knowledge and network accumulated during their experience as fund managers in the wind industry proved to be beneficial for them to start an enterprise.

To summarize the research gaps identified by the previous authors, the current study of the network model, the digital ecosystem model and the internationalization theory needs to be further reinforced with the reality of Chinese tech firms. Therefore, with the study conducted on the smart health industry in China, the author of this book hopes to expand and extend the international business theory.

3. Theoretical Framework and Research Methodology

Chinese tech firms have joined the top tier manufacturers in the Internet of Healthcare Things industry. After 20 years of industrial policy focusing on upgrading Chinese firms on the value chain, the policy has finally started to pay off. The Chinese government promulgated 'Made in China 2025' and 'Healthy China 2030', projecting the use of big data and cloud computing to upgrade the Chinese healthcare system and to counteract the effects of aging population.

Take the value chain for wearables as an example—it can be roughly divided into 8 parts (Wearable Technology 2019):

1. Chip vendors (component or material): For instance, companies such as Texas Instruments, 3M, NXP, Bosch, Intel and Qualcomm are categorized as chip vendors (components and materials) (Wearable Technology 2019). Companies such as Alibaba and Huawei have recently started developing their own chips called Hanguang and Ryzen. Companies such as Texas Instruments, Intel and NXP design and manufacture semiconductors from silicon (Intel 2020, NXP 2020); NXP also provides IoT and automobile solutions based on semiconductors. 3M deals in soft materials, equipment and machines for product manufacturing (3M 2020). Qualcomm has accumulated a large number of patents in chip design (Qualcomm 2020). Bosch produces sensors which can be incorporated into the IoT system (Bosch Global 2020).

2. Electronics Manufacturing Service (EMS), design, embedded systems, integrators, Original Equipment Manufacturers (OEMs), Original Design Manufacturers (ODMs): Companies like Accenture, Foxconn and Cosmo Supply Lab belong to this category (Wearable Technology 2019). Accenture serves as a consulting firm which focuses on designing smart ecosystems (Accenture 2020). Foxconn's business

lies in assembling different parts in smartphones and other wearables (Foxconn 2020).

3. Standardization Test Houses: Huawei has now become an important part of standard setting, with research staff volunteering for IEEE; the recent turbulences caused by the trade war even led IEEE to ban Huawei staff members from participating as reviewers and editors. Zigbee, Wifi, Bluetooth and other standardization testing houses are also related to the development of the IoT industry (Wearable Technology 2019). More and more Chinese firms are actively participating in the standard setting process for technological development. For facial recognition standard setting, for instance, Chinese firms like Sensetime are leading a consortium of tech companies in setting global standards at the UN (Financial Times 2019). This has caused a panic in the US as the US government believes that Chinese firms are taking a lead in setting global technology standards for facial recognition. Subsequently, CIFUS banned US companies from cooperating with Chinese companies such as Sensetime and Yitu on national security grounds.

4. Network providers: In China, state-owned network providers dominate the market. For national security reasons, China Telecom, China Mobile and China Unicom dominate the broadband internet and mobile service market. There are 1.578 billion individual mobile users with 113.8% market penetration rate in China (World Bank 2019). Among the 1.5 billion users, China Mobile accounts for about 9.35 billion, and the three service dominators take up to 99% of the mobile service (data, calls and text message) in China (GSMA 2020). In Europe, Telecom, Vodafone, O2 and Wind offer internet and mobile data services. In the US, companies such as Verizon and AT& T take the lead in providing mobile services.

5. Cloud Services: Alibaba, for instance, has become a strong player in the field of cloud storage and computing. It provides both private and public cloud services in data storage, computing, algorithm optimization and other related services. Google Cloud and Microsoft 365 offer similar solutions. Alibaba is selling its cloud services to its ecosystem partners in South East Asia (Financial Times 2020), such as Lazard, to expand the market share of its cloud services. Tech companies like Xiaomi have data stored in Singapore, the US and China, depending on the local laws (Xiaomi 2020). Baidu as a tech company also offers cloud solutions for both individual users and institutional customers (Baidu 2020); Huawei offers similar services as well (Huawei 2020). Almost all significant Chinese tech companies offer such services and subsidize individual users significantly with

free services in the hope of using the vast amount of data to train their algorithms.

6. B2B & B2C services providers: Nike, Runtastic, Apple, ING, Mastercard, Google Pay, Alipay, ING and Paypal fit into this category. Nike and Rubtastic (sponsored by Adidas) offer personalized training advice, while Apple iOS offers a digital platform for app developers, and ING, ABN AMRO, Alipay, Google Pay and Paypal offer digital payment services.

7. Product solutions: Huawei, Philips, Huami, Garmin, HTC and Adidas belong to this category. Huawei, Philips, Huami, Garmin, HTC and Adidas provide users with wearables, phones, shoes, medical devices, etc. Therefore, these companies are product focused companies.

8. Distribution: Apple Store, Amazon, Bestbuy, Huawei Flagship Store and Xiaomi Flagship Store fall under this category. Retailers are the first points of contact of users and brands. Hence, distribution plays a significant role for brands to know customer preferences and accommodate to the customers' needs by adjusting their brand image. The theoretical framework for this was established by Wearable Technology (2019) to evaluate the position of firms in the value chain. Alibaba started from the bottom of the value chain as distributors; then it moved up to being product solutions providers, service providers, cloud service providers and ultimately chip venders. Huawei started as product solutions providers, then moved up to being service providers, network providers and standardization test house.

To summarize the above-mentioned factors, there are unanswered questions about factors affecting the telehealth industry development and how Chinese tech companies upgrade their positions in the value chain of the telehealth industry within a short period of time. To answer the research question for the book, case studies based on the Chinese smart health industry have been presented below.

To answer the research questions described in the previous paragraph, the case study was designed with guidance from Yin (2017) and Eisenhardt (2012), to study the ecosystems of tech companies and their incentives in exploring the digital health industry. Three companies were considered for the study—Huawei, Xiaomi and Alibaba.

Semi-structured interviews were held with the Business Manager of Alihealth—Medical Brain Unit, the Business Manager of Huawei Wearables and Sports Health Unit, the Business Manager of Xiaomi Wearables Unit and the Investment Manager of MIUI Investment Division, in an anonymous manner.

These interviews were aimed at getting insights from tech companies involved in barriers and strategies for implementing smart health solutions in China. The data was collected in Chinese, translated by the

Figure 2.3: Wearable Technology Value Chain. Source: Wearable Technology 2019; Author's design.

same researcher in English, and consent was obtained for publication and quotation anonymously.

The interview questions were designed and sent to the interviewees in advance. If time allowed, additional questions were proposed to get the questions answered thoroughly. The questionnaires are attached as appendix to this chapter.

4. Innovative Business Model Case Study based on Huawei, Xiaomi, Alihealth and ByteDance

The smart health industry in China is a burgeoning market which offers opportunities of significant chances of growth to foreign and domestic tech companies alike. Table 2.1 lists start-ups in China and in Germany, offering services such as improving diagnosis efficiency, proactive health management and personalized medicine. The start-ups have potential to upgrade the smart health industry with their technology, business models and digital network.

Table 2.1: Services offered by selected smart health start-ups. Source: Sensetime 2020, Yitu Technology 2020, Linkdoc 2020, Keep 2020, Burning Rock 2020, Mobvoi 2020, Inveox 2019, Wegene 2020, Infervision 2020.

	Cloud Computing Platform for Developers	AI appliances in Diagnosis	Personalized medicine	Genetic Test	Internet Hospital	Wearables	Personalized Training Program	Partnership with insurance companies	Business Model
Sensetime	X	X							B2B
Yitu Technology	X	X							B2B
LinkDoc					X			X	B2B
Keep							X		B2C
Burning Rock Dx			X	X					B2B & B2C
Mobvoi						X			B2C
Inveox		X							B2B
Wegene				X				X	B2C
Infervision		X							B2B

Sensetime and Yiyu are tech start-ups which work on image recognition. Mobvoi provides a voice assistant with its wearable products in different user settings. Linkdoc works in the domain of internet-based hospital solutions and offers AI support for pharmaceutical companies. Burning rock aims at identifying genetic linkages with cancer and provides personalized medicine solutions.

As mentioned earlier, the interviews for this study were conducted with Huawei, Xiaomi and Alihealth. Other companies were studied based on their public profile—such as company website, news and public information disclosure (quarterly and annually) if they were listed in the stock market. In Table 2.2, Huawei and Wedoctor have been stated as privately owned companies, given their positions in digital infrastructure, hardware and software. Tencent, Ping An, Xiaomi, Alihealth, Microsoft and Samsung are public listed firms. We Doctors and Tencent Clinics offer services which can be partially reimbursed by the basic employee medical schemes and some private medical insurance schemes. Tencent Clinics, We Doctors, Ping An, Alihealth and Samsung have collaborative relationships with private insurance companies. The partnership implies that users can be diverted from the abovementioned service platforms to insurance schemes, or may offer collaborative service packages with fellow insurance companies so that insurance companies may dive into health-related data while users get more services and less choices from the same package.

Table 2.2: Financial Ownership and business network of other Chinese firms in the telehealth industry.

Competitor	Public	Private	Pension Systems Integration	Partnership with Insurance Companies
We Doctors Limited Holdings		X	X	X
Tencent	X		X	X
PingAn Healthcare and Technology Ltd	X			X
Huawei		X		
Xiaomi	X			
Alihealth (Alibaba)	X			X
Microsoft	X			
Samsung	X			X

4.1 Case Study 1: Huawei

Huawei as a firm has received wide public attention lately because of the US—China trade war (WSJ 2020, BBC 2020, Financial Times 2020,

Bloomberg 2020). Huawei has obtained a portion of the global market share in IT network infrastructure, which attracts attention in the upgrade to 5G, given the national security concerns of western governments (see Figure 2.5). For the smartphone category, Huawei has also obtained a considerable market share (see Figure 2.6). Huawei's ecosystem also incorporates wearables (bands, watches), apps, smartphones and home appliances for individual users. In the smart health industry, for institutional users, Huawei offers e-Hospital solutions, regional health connection solutions, regional healthcare information network solutions, telemedicine, and multi-channel HD telemedicine solutions. These solutions were used during COVID-19 for hospitals in Wuhan to enable tele-conferencing with experts from other provinces, to call doctors from the designated hospitals, hosting about 700 patients with COVID-19.

Huawei has a wide customer base, including children, working professionals and institutions. Their sales channels include online, offline, and social media campaigns via Weibo. Over the past 20 years, Huawei has risen from being merely a Chinese brand to become an international brand (it was created to sell switches to telecom service suppliers initially).

Huawei follows a hardware-based strategy; it started with switches for telecommunication in the 1990s. During the process, Huawei was almost obtained by Motorola; but later, it started to etch in from the top of the supply chain to the bottom to supply chain. Huawei's investments in R&D and human resources are huge, which is probably why it never went public. In 2018, Huawei devoted RMB 101.5 billion (USD 14.48 billion) to R&D, accounting for 14.1% percent of its annual gross income (Xinhua Newsnet 2019). It also obtained 5405 patents (Xinhua Newsnet 2019). The solid redemption packages offered by Huawei attract a large number of young as well as old recruits to join the firm. The youngest VP in Huawei was 27 when he was promoted for solving major technical challenges such as developing new types of switches. In the beginning, the competition was fierce between Huawei and other major telecommunication equipment suppliers like Bell.

Huawei has expanded globally at a remarkable pace due to the fact that it sent its staff to areas such as Uganda as well as Greece—markets which were previously ignored by western firms like Nokia, Ericsson, Qualcomm, etc. These markets are closely aligned with the low human resources cost and production cost of Huawei.

Since Huawei has obtained a large market share in the upstream of the value chain in the ICT industry (such as IT standard setting), it has expanded into the downstream of the value chain as well. For instance, Huawei has acquired a large market share in the smartphone industry (see Figure 2.6). However, as the U.S. government imposed an export ban on Huawei, Huawei separated the 'Honor' brand, essentially dropping

TE30 All-in-One HD
Videoconferencing
Endpoint

TP3206-55 Panovision
Telepresence System

Figure 2.4: Huawei Ecosystem Display for individual users and institutional users. Source: Huawei Consumer Wearables 2018, Huawei Healthcare Solutions 2018, Developers Huawei 2018, Research and Innovation, Huawei Europe 2018.

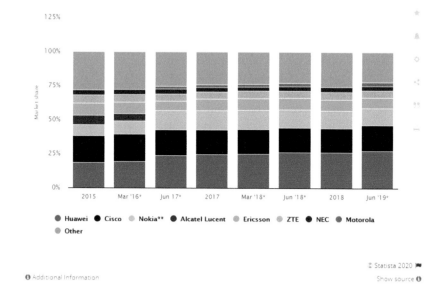

Figure 2.5: Market Share of IT Network Vendors, 2015–2019. Source: Statista 2020.

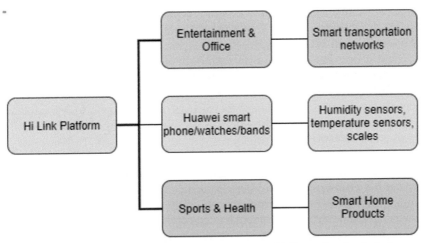

Figure 2.6: Huawei IoT ecosystem. Source: Interview with Huawei, author's design.

the lower end consumer business to a separate brand to maintain its dominance in the enterprise business.

Huawei has also expanded into the downstream of the smart health industry, such as wearables. Despite claims of having no health focused strategy, Huawei has adopted a combined software-plus-hardware approach to tackle healthcare problems. For hardware, they offer Huawei smartphones, wearables and IoT-related smart home products such as sleep monitors (non-medical devices), scales and masks. These products are sold via two channels—online and offline. The online channel includes Vmall, Taobao, Tmall, Amazon, Weibo, etc., and social media campaigns. The offline channel includes Huawei flagship shops abroad and in the home country. It has been estimated that, in 2018, Huawei had about 2000 offline flagship shops in major shopping malls in China. In Europe, Huawei has set up experience centers in Paris, London, Brussels and other major cities.

What is worth mentioning is Huawei's pricing strategy. In the beginning, Huawei was seen as a technical solution provider to institutional users. Over 36 years, it has established a solid individual user base in China as well as overseas as a reliable brand—among children, professionals and middle-aged user groups alike. The brand image has switched from a low-end brand to a high-end brand capable of competing with Apple and Samsung in the smartphone market. In the overseas Chinese market, Huawei focuses on selling smartphones and wearables as packages to promote the use of their wearables. For instance, the smart bands offered by Huawei has a price range of RMB 129 to RMB 459 (Euro 17.9 to 61.19) in China; the Huawei smart watches range from RMB 449 to 4988 (Euro 59.86 – 665) in price; the smart sleep monitor costs about

RMB 999 (Euro 133.08), while the smart scale costs around RMB 113 (Euro 15.05); furthermore, their smart masks cost around RMB 249 (Euro 32) (Huawei Vmall 2018).

In Europe, Huawei has a staff strength of about 1900 in R&D. To make use of the European innovation capacity, Huawei had established about 18 R&D centers in Europe by 2018 (Huawei 2018). For instance, Huawei has an R&D center in Germany because of its industrial design capacity and has since promoted the Huawei Mate Porsch series phones for the high-end market (Huawei 2018).

For the internet of healthcare things, Huawei tries to develop the ecosystem with its partners on the platform. The IoT smart home products under the Huawei brand involve humidity and temperature sensors and smart scales. The other products available on Vmall are developed by the ecosystem partners. Huawei has created a connecting platform called Hi-link to connect more than 50 different kinds of IoT products covering 4 categories: sports and health, entertainment and office, connected cars, and smart home products. Wearables targeting the aging population are still in the development stage; the possible functions therein include fall detection, humidity and temperature sensors, GPS, and heart rate and blood pressure monitoring. Regarding data sharing within the user group, Huawei has explicitly pointed out the need to obtain user consent.

Huawei aims to offer preventive healthcare services as well as post-treatment and long-distance patient care in the future. These solutions include developing algorithms for automatic nutrition detection on smartphone cameras, and edge-based computing for health-related data (weight, heart rate, sleep, sports, etc.). Huawei is currently working to develop its own algorithm and chips for edge computing. Huawei has also explored a data-sharing based business model by working with an insurance company in South Africa, serving users with free wearables to give them incentives for exercising, with adjustable premiums.

Huawei has identified certain barriers in implementing smart health solutions. For instance, no separate technology admission standards exist for the application of wearables in the medical sphere in China. Also, so far, no patient profile platform exists at the national level either. Furthermore, there are culture-related barriers for marketing and developing user-based functions in Europe. Huawei offers a basic tracking function for wearable products being sold in Europe, limited third-party partners such as internet-based hospitals.

Going into the details of the Huawei ecosystem, Huawei uses phones, tablets, sports watches, apps (Huawei Sports and Hlink) and routers to interact with its users. The Huawei Hi Link App has been designed for smart application scenarios in home entertainment, home lighting, smart health, energy management, home security, automation, routers, smart TV boxes, lighting, environment, interior design, all coverage routers,

Wifi, kitchen, audio, smart sleep monitoring, etc. There are about 50 core partners.

Huawei has promoted the use of interaction depending on the maturity of the technology. Now Huawei is focused on delving into the health-related function of wearables and terminals (telephones). For instance, there are temperature sensors on phones, which can detect temperature inside and outside the house in the future. The same design is planned for smart watches. Meanwhile, there are temperature and humidity sensors on smart home devices. Huawei is working with third party partners such as Omron and Dnurses to track blood pressure and blood sugar level changes. Huawei mainly uses its terminals to keep track of the biometic indicators.

In China, Huawei smartphones are pre-installed with Wedoctor, DXY and Chun Yu Doctor apps to make it convenient for users and to provide users with better healthcare experience and offer one-stop health services. In the overseas market, such as Europe and the US, wearable devices manufactured by Chinese companies face strict regulation. Meanwhile, to develop health-related functions, it is necessary to consider the local culture. Therefore, it is not possible to delve in the value for wearables compared with the Chinese market. The development process is oriented towards users in the Chinese market. There are a few restrictions to launch the sleep and sports related functions in the overseas market as well. These functions are not going into the medical directions yet. We are yet to see whether there will be restraints towards wearables and medical related embedded functions in the future.

At present, the South Africa insurance company, Discovery, is working with Huawei to offer free wearable devices, charging users on a term basis. When users reach a certain number of steps daily, the monthly term payment can be exempted. Users can lower their insurance premium by using the device for long terms; in the meantime, it is possible to lower health risks for users.

As a multinational, Huawei Headquarters in Shenzhen took up the function of coordination of resources. Their wearables are mainly developed in the domestic market. The Huawei HQ in China is a coordination center to manage all the businesses for Huawei growth overseas. The location for the overseas development center was chosen based on technology, like the algorithm development center in Finland, the hardware development center in Germany and the design center in France (for fashion), etc. The sample in the labs for developing algorithms is relatively small. It is mainly developed in labs, does not use data from users, and will not transfer data from China to overseas labs or vice versa. It is possible to update the algorithms later at the clinical trial or commercialization stages with larger size of data.

4.2 Case Study 2: Xiaomi

Xiaomi has become a leader in the Chinese IoT market. Xiaomi also produces Mi Band, a smart health device which offers coaching advices to users in different scenarios.

Xiaomi started off as a smartphone manufacturer in 2010 in Beijing. Founded by Leijun, an entrepreneur who had multiple years of experience as the CEO of a software corporation, Jinshan, which specializes in offering WPS (a Chinese version of Office) solutions. By 2018, Xiaomi had become the world's leading IoT device manufacturer. In China, Xiaomi's marketing position has been low-end and middle class. Therefore, their major marketing campaigns focus on young people and students. In the overseas market, Xiaomi targets the middle class and high-end market.

Xiaomi's products cover several categories: wearables, which include bands, watches and shoes (customization/prosumer design); home appliances, which include smart scales, smart air purifiers, smart toothbrushes/lights/water purifiers, etc.; and smartphones, which include Red Mi targeting young users, smartphones for pets, elderly and children, with only GPS tracking and voice communication functions. Xiaomi's market share is the world's second largest by shipment for bands.

Figure 2.9 illustrates the Xiaomi IoT corporate governance structure and business model.

Figure 2.9 displays Xiaomi's IoT business model.

Xiaomi has created partnerships with insurance companies for two purposes: to increase the user adherence to Mi Band and to promote the products in a separate user group by diverting users from Mi Fit to Zhong An Insurance for Steps Sharing Insurance. In this way, users can trade steps walking per day with the day insurance premium payment to Zhong An. However, the partnership stopped because of GDPR and the P.R. China Privacy Law (Chen 2022).

To build the IoT product matrix efficiently at a low cost, Xiaomi has adopted the following strategies.

Xiaomi focuses on 80% of its users and their 80% of the needs, instead of focusing on the entire user base. It reduces costs and offers standardized products. Therefore, Xiaomi users do not have the luxury of personalization of products. The Xiaomi brand is seen as a low-end and middle-class brand targeting the young user group in China, compared to Huawei's more sophisticated brand image. Xiaomi focuses on core products and strategy design. The ecosystem brands are responsible the new dimension in the IoT matrix (Chen 2022).

Xiaomi has also identified barriers of going into the IoHT market as data sharing with doctors when it comes to healthcare data and personalized healthcare solutions can be difficult in China, given the scarce

Company	1Q22 Shipments	1Q22 Market Share	1Q21 Shipments	1Q21 Market Share	Year-Over-Year Growth
1. Apple	32.1	30.5%	30.1	27.7%	6.6%
2. Samsung	10.9	10.3%	12.1	11.1%	-9.9%
3. Xiaomi	9.8	9.3%	12.9	11.9%	-23.8%
4. Huawei	7.7	7.3%	8.6	7.9%	-10.8%
5. Imagine Marketing	3.2	3.0%	3.0	2.8%	5.2%
Others	41.7	39.6%	41.9	38.6%	-0.5%
TOTAL	105.3	100.0%	108.6	100.0%	-3.0%

Figure 2.7: Market share of wearables unit shipments worldwide by vendor from 1Q'21 to 1Q'22. Source: IDC Research 2022.

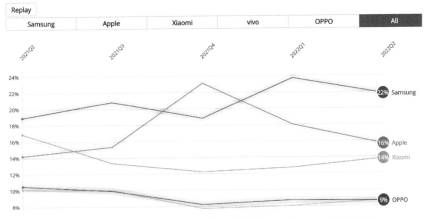

Figure 2.8: Smartphone market share, IDC Report, Source: IDC 2022.

medical resources in the country. The CFDA processes can slow down the development of new health-related functions. Xiaomi is conservative regarding sharing data collected from wearables with others. Besides Xiaomi, there have been a few attempts to test the data driven business models for smart health solutions.

For example, users share steps data with insurance companies to trade their steps for premiums of critical disease insurance. Others can trade steps for term payments of wearables. Some companies decide to rent smart home devices and offer coaching advices for the elderly. Users pay for services for fixed terms and get the device for free in the end.

Xiaomi users are getting used to paid services with IoT products serving their needs, but it is not the time to harvest its value yet. The current investment project is mainly focused on complementing Xiaomi's existing internet-based businesses, to create synergy value, or to fill in the gap for up/downstream businesses. Xiaomi has invested in

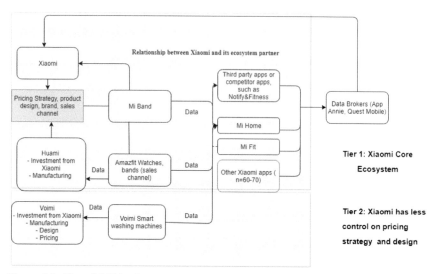

Figure 2.9: Xiaomi IoT business strategy. Source: Author's design, Interview with MIUI Investment Director.

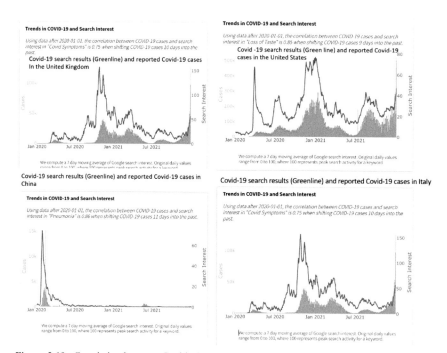

Figure 2.10: Correlation between Covid-19 and search trends of Covid 19 symptoms in the U.S., U.K., China and Italy Source: World Bank Data Analytics, 2022.

51Yund.com,[2] Ximalaya.com,[3] Ofo,[4] etc. Now Xiaomi offers internet services as well—mainly advertising and Internet value-added services,[5] currently accounting for 9% of its annual revenue, Investing in these apps, which come with a large user base, can potentially complement Xiaomi's current internet services as Xiaomi cannot offer all these services on their smartphones. Xiaomi now has a large user group with users demanding a vast number of services. The unmet needs of users can be met with the apps Xiaomi has invested in. Xiaomi also offers funding and platform to start-ups; as these businesses grow, it can be profitable for Xiaomi to get returns from the ecosystem businesses and by diverting users from Xiaomi to these apps. Their business model is about diverting users or to complement services for Xiaomi. The value of the business model lies in complementing Xiaomi's services, creating synchronized value and eventually profiting from shareholding of ecosystem firms.

In India, Xiaomi has a localized brand image with Manu Kumar Jain nominated as Head of Xiaomi India. This move was aimed at impressing the Indian consumers with the Xiaomi brand and assuming a high-end brand image as Manu Kumar Jain is seen as an aristocrat in India. Xiaomi remains popular with young users in India, with the large android market offering a wide choice of apps to analyze data collected from Mi wearables. This seems to be a common marketing strategy in Asia—while middle-class brands in Korea and Japan become high-end brands in China, middle-class brands in China become high-end brands in India.

4.3 The use of smart health solutions in monitoring epidemics

From a public health perspective, whether Google can serve as a reliable source of public health information is questionable. Nonetheless, Google remained a public source of getting information regarding the then ongoing Covid-19 pandemic. Google uses its strong analytical tools when it comes to search results and search trends. By analyzing search trends, the public keeps track on the transmission of Covid-19 variants (Google Trends 2022). For predicting the trends of Covid, Google search trends proved to be highly correlated with Covid-19 cases (Pan et al. 2020). Pan et al. (2020) noticed one important difference between China and the rest of the world, with the predictability of Google Trends for Covid-19 outbreaks. The reason for such a phenomenon is that people in China generally use Baidu, Alipay, Zhihu, Jinri Toutiao and Wechat, instead of

---

[2] 51Yund.com offers sports coaching services.
[3] Ximalaya.com offers music, audio books and educational services.
[4] Ofo.com offers bike sharing services.
[5] Xiaomi.com, http://blog.mi.com/en/2018/08/06/the-internet-reimagined-xiaomi-provides-tailored-internet-services-to-meet-evolving-consumer-needs/.

Google, to access public healthcare information. The World Bank carried out a correlation analysis between Covid-19 and the search trends for Covid-19 symptoms—such as loss of taste, loss of smell, cough, fever and shortness of breath from January 2020 to January 2021; strong correlation was found in the U.S., U.K., China and Italy, as indicated by Figure 2.11.

Other literatures suggest that Google Trends for vaccination have been used to predict the vaccination rate in Italy (Rovetta 2022). It has been suggested that the government uses Google Trends as a complementary tool to predict and monitor the vaccination distributions in Italy.

Google had used smart health solutions to monitor epidemics such as flu back in 2013. Back then, Google Health API monitored flu trends by the number of search results (standard deviation from baseline) in several countries. The API allowed public health experts to monitor the rise and fall of flu cases in the country, and selected and compared the trends between different healthcare systems. Take the Netherlands, Germany and Belgium for instance. One can observe from Figure 2.11 that the flu search results spiked from January to February every year except 2009 (where the peaks appeared in both January 2009 and November 2009). This can be explained by the H1N1 epidemic in 2009. The differences in intensities of search activities can possibly be explained by the variance in the responsiveness of the public health systems in the Netherlands, Belgium and Germany. Another factor which plays a crucial role is the attitude of the healthcare system towards epidemics like flu. In Belgium, flu patients are advised to seek consultation and get subscription from doctors, while patients are often advised to stay at home and perform self-enforced quarantine in the Netherlands and in Germany. The last factor which may explain the difference in search activities can be the varying levels of trust in public healthcare systems in Belgium, the Netherlands and Germany. The lack of trust in government organizations can be explained through the higher levels of hierarchy and the lack of transparency.

With the COVID-19 outbreak across the globe, by Nov 16th 2022, total infection in China had reached 9,258,687 confirmed cases and 29,112 deaths caused by COVID-19 were reported in China (WHO 2022). The use of smart health solutions can be observed in epidemic monitoring, online consultation and CT image screening to improve diagnosis efficiency, medication sale and payment. The outbreak of COVID-19 has promoted the use of telehealth solutions by the government, the private sector, as well as the general population, for the prevention, control and monitoring of the pandemic. Whether the trust in telehealth solutions has increased in China is yet to be tested. However, with the Zero-Covid policy lockdowns, many patients had no other option but to consult doctors online through Alihealth, JD Health, Wedoctor, Xiaohe Health, Ping An Good Doctor, Chunyu Doctor, etc.; whether the online consultation service remains reliable or not often depends on the quality of doctors; the paradox here

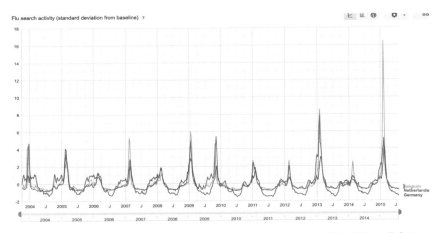

Figure 2.11: Flu search activity—standard deviation from baseline, 2004–2015, in Belgium, the Netherlands and Germany. Source: Google Inc.

Figure 2.12: Global COVID 19 cases by November 16th 2022. Source: Johns Hopkins University Corona Virus Resource Center 2022.

is that good doctors who work in level 3 hospitals in China are often very busy and do not have the time for online consultation; the doctors who do have the time for online consultation are doctors who are not trusted by patients for in-person consultations (Deloitte 2020). There are several hospitals which offer online consultation services to returning patients in case of chronic diseases, and offer online pharmaceutical services as well; one such example is the Union Hospital in China, which offers a full set of services—from online consultation to online pharmacy—for vulnerable groups of patients who are afraid of going to hospitals.

In the beginning of the pandemic, because of the government-enforced quarantines, residents were forced to stay at home. They thus used online resources to track the epidemic and take precautionary measures. Going to supermarkets, hospitals and pharmacies were seen as risky behaviors or were not allowed at all. Residents in China therefore turned to online platforms to shop for groceries and order medication, masks and goggles. Because of the quarantine, people were encouraged to report their sickness via government controlled apps and Wechat mini programs, apply for travel permits, get masks, report price distortions, and search for real-time information regarding COVID-19.

With almost three years in the global pandemic, despite the huge amounts of financial resources put into pandemic control, China still stuck to a dynamic zero-covid policy. The reasons were the low vaccination rates among the elderly residents and the fear of COVID variants. While the rest of the world chose a prevention-based COVID fighting strategy with m-RNA vaccines, the Chinese government adhered to Sinovac (developed by the national pharmaceutical company). The longer the zero-Covid policy stays, the less is the willingness of vulnerable groups to be vaccinated.

The fear of loss of control over the outbreak overwhelming the healthcare system has risen as COVID testing and control drained the public healthcare system of their financial resources, particularly personnel resources. This delayed China's reopening to the rest of the World. Different local governments looked at the target of zero-Covid as a political goal. The zero-Covid policy became less public health related as it turned into more of a means for tracking Chinese citizens and controlling the Chinese population from traveling inside China as well as between China and overseas. The goal was to control all the intermittent contacts for anyone with Covid-positive patients, and not the cost to the healthcare system.

The restrictions on travelling prevented China from emerging from the pandemic-induced economic recession, also leading China to multiple macroeconomic shocks such as the real estate sector crisis, drop in consumer demand, lack of confidence for entrepreneurs, etc.

Figure 2.12 and Figure 2.13 present two examples of how apps can be used to monitor the epidemic. Figure 2.12 displays the real time monitoring of global COVID cases (developed by John Hopkins University). In the case of Italy, there were around 31,506 confirmed cases of COVID-19 by March 18th 2020 (John Hopkins University 2020).

Figure 2.13 shows a real-time map of COVID-19 cases in China. In the top search bar, one can input the name of a city and get the number of confirmed COVID cases. The figure shows that in Xiangyang, Hubei Province, the search results display a real-time number of 1,175 confirmed COVID-19 cases; it also informs the general public that all confirmed cases have been treated at the designated hospital and reassures them to stay at home to avoid further public panic.

There were complaints from the western society and panic fuming as people started to accumulate groceries such as toilet papers at the

- Location based Covid tracking

- There are currently 1175 confirmed cases in
- The city you selected: Hubei Xiangyang. The cases
- have been admitted to designated hospitals.
- Don't worry so much and take care of your self.

- The warm warning from Alipay

- Wear the mask, wash your hands, do not gather socially
- and survive the global pandemic together.
- Statistics until March 3rd 2020, 03:28:59

Figure 2.13: Real-time COVID-19 cases in Xiangyang, Hubei, China (right), English translation (left) by March 3rd 2020. Source: Alipay 2020.

beginning of the pandemic. The panic could have been caused by the lack of trust in the data about coronavirus as disclosed by the government. In this pandemic, there were major issues of the public healthcare information disclosed by the government and accepted by the public.

Firstly, when cases were first detected in the wet markets in Hubei, no immediate quarantine methods were taken for the patients. The layers of beaucracy between the local health commission and the National Health Commission slowed down information reporting. The local health commission, knowingly hide the cases from the public until Spring Festival Break, trying to pretend nothing serious has happened. This caused national and international spread of the disease at a later stage. China is not new to SARS, and pandemic control/prevention. For Chinese the citizens, the number of COVID-19 cases were disclosed rather late, with risks unclarified in the beginning.

Secondly, the funding cuts for public health spending and on local disease control center suggests that the traditional public health control system gradually stopped functioning whereas the general public trust information from public sources. The private sector actors, private capital and tech companies play a much larger role in public health related information dissemination and control.

Thirdly the public, failing to receive sufficient public health knowledge regarding vaccines, Covid prevention and control, did not have a choice when it comes to differ between information/disinformation. The subsequent lockdowns and travel restrictions are increasingly bewildering even for public health experts. Fighting over Covid became ideological, with political goals, aims and targets mixing with scientific facts. The Chinese government knowingly chose a less effective vaccination, instead of importing 'foreign' m-RNA based vaccines, to protect domestic state-owned pharmaceutical companies. Compared with other healthcare systems which took a more practical approach, the Chinese government took a more systematic control over the population over the cause of Covid prevention. Local governments over spent on quarantine centers out of budgets. The central government put the number of Covid cases as top priority.

Fourthly, the three years of lockdowns have significant implications economic implications. With all the pessimistic deleveraging process going on, the rhetoric supporting the Covid policies are being doubted and questioned with more and more Chinese vote with their feet (WSJ 2023). The housing market crackdown, the bankruptcy of small-scale real estate firms, the slowdown of real estate projects, and the loss of confidence of house buyers have caused a downward spiral with dreadful economic consequences. China, which once sits at the center of global supply chain now faced on the one hand global political fractions over how the government handled Covid, and on the other hand, lost the economic propellants which sustained the economy and confidence for two decades. Domestic demand has diminished as consumers in general have lost confidence due to uncertain economic future. The value of relevant financial investments, such as a second home, securities, bonds, trust products have dropped. Businesses, faced with uncertain geopolitical environment and constant disruptions for supply chains, are seeking alternatives outside China; of course, some are investing in factories in Vietnam, India, and other South East Asia facilities instead of China.

In the United States, public health institutions experience a similar dilemma. As public health data and information was spread not through the traditional media channels, but through multiple social media platforms such as Twitter, Instagram, Facebook, it is difficult for the general public to discern information from disinformation. Multiple issues such as whether to wear a mask or not, whether to take the vaccination or not, and which vaccination to take, whether social distancing is necessary, flexible working schedules, have all become political and debatable.

To summarize, sovereign states may believe public health information spread may follow the traditional manner (as indicating by Figure 2.14a). However, in reality, information flows follows an alternative pattern where the tech companies stand as independent entities apart from the sovereign state; interactions between the public and tech companies are more often than the sovereign states are used to (Figure 2.14b).

In Europe, there were reasons for citizens to believe that the public healthcare system reacted slowly and rather irresponsibly to control the spread of the disease. In China and Italy, when it comes to

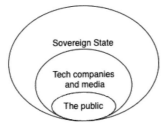

Figure 2.14a: Traditional Information Flow. Source: Author's illustration.

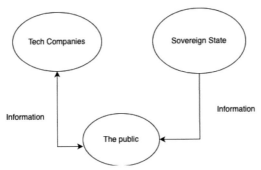

Figure 2.14b: Information flow during the Covid-19 pandemic. Source: Author's illustration.

disclosing public information, both the governments (whether at a local or the national level) showed a number of inconsistencies. The Chinese government thus learned from its own mistakes and changed the head of the Hubei Province Health Commission and Wuhan Municipality. Public information disclosure was made via apps subsequently. After a national level quarantine, working remotely was widespread in China, with major satellite maps showing a major decrease in nitrogen related pollution (NASA 2020). The level of industrial activity was also shown to have decreased significantly. The enforced quarantine caused wide downward economic projections around the globe. The European Commission was hence taking measures to reduce the exposure of the economy to COVID-19 (European Commission 2020).

Figure 2.15 shows the free public consultation service offered by Level 3 hospitals in Shanghai, e.g., Ruijin Hospital affiliated with Shanghai Jiaotong University, Shanghai Public Health Center Clinical Center, Shanghai No.1 Maternal and Infant Care Center, Shanghai Chinese Medicine Hospital Yangpu District, Shanghai International Medical Center, Yueyang Chinese Medicine and Western Medicine Center, etc. The consultation service offered through chat (text and picture) with the doctors is free.

There are also reports from Italy indicating the use of such methods in the country, the effectiveness of which cannot be judged as ICU beds

Online consultation during Covid lockdowns

Working together to combat Covid

- Shanghai Public Health Clinics Center

- Shanghai First Maternity and Infant Hospital

- Shanghai Jiaotong University School of Medicine

- Yueyang Hospital of Integrated Traditional Chinese and Western Medicine Out-patient Department

- Shanghai Yangpu District Chinese Traditional Medicine Hospital

- Shanghai International Medical Center

Figure 2.15: Shanghai Hospital Online Consultation (left, right with English translation) by March 3rd 2020. Source: Alipay 2020.

in Italian hospitals are not enough for patients in need of intensive care (Grasselli et al. 2020). With a significant aging population, the death toll from COVID-19 was projected to be higher for Italy than for China.

4.4 ByteDance and its business model

In 2020, Bytedance, the company behind TikTok, set up a big health department—with the name Northern Light. The Northern Light department covered several parts of the business—internet hospital, medical information, AI aiming to assist drug development, medical education for the general public, offline clinics ('Pinecone Clinic') and general practitioners.

ByteDance was successful because of TikTok; the short-video sharing platform built a reputation among young users domestically and overseas. After this social media platform became extremely successful, ByteDance sought alternative growth paths in online education, online reading, online gaming, and other content streaming and sharing platforms. However, with the Chinese government cracking down on the tech sector, especially on online education platforms, the healthcare sector seemed worth investing in. In Dec 2020, Bytedance acquired Xiaohe Medical, an internet hospital (CB Insights 2020). Yet, compared to Alibaba and JD Health, for which a large part of the business is focused on e-commerce, the business model of Xiaohe Health-subsidiary of Bytedance, focuses more on helping patients build communities and self-help groups, offering reliable information to patients, developing AI-based diagnostic software as well as pharmaceutical R&D, and offering offline clinical services.

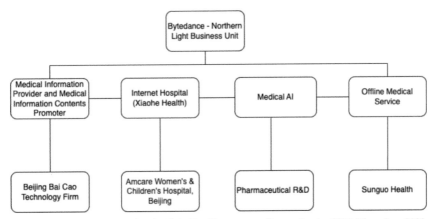

Figure 2.16: ByteDance entering the digital healthcare space, Source: Levine 2022, Bloomberg 2022.

The ByteDance business model shows how a short video app can spread globally in a relatively short period of time (see Figure 2.16). ByteDance was founded in China in 2012, with the launch of the TikTok app, and has now accumulated over 400 million daily active users. However, in India, the company has had trouble with the contents on the referred app. TikTok had obtained about 120 million users in India. The Indian government had then decided to temporarily ban the app in the app stores of iOS and Google Play, following the judgement of Madras High Court (Findlay 2019). As a company whose main profits came from advertisement and targeted marketing, the app ban posed a major threat to the ByteDance business model. In order to adapt to the challenge, ByteDance undertook various measures to adjust its market strategy in India within 3 months of the ban. After the flood and tornado hit India, ByteDance sent its Public Relations and Corporate Social Responsibility Team to the country, delivering poverty reduction and disaster resolution materials. In addition to such humanitarian efforts, ByteDance had tried to set up a data center in India to smoothen its relationship with the Indian government. Moreover, the company had taken an education initiative, launching an English learning app 'Gogokid' and a learning app called 'Haohaoxuexi'. Realizing the potential in education tech market, TikTok had started to collaborate with the local partners. It had worked with 'Testbook', an app which helps test candidates to prepare for exams. TikTok had adopted this strategy for two purposes: (a) to create more use cases by diverting the business from mere entertainment, getting to the core needs of the users, and increasing user stickiness; and (b) to win the trust of the local Indian market, including regulators. Since collaborating with TikTok, 'Testbook' has around 10 million users diverted (Findlay 2019). To summarize, with a large user base, and customer acting as an important partner for TikTok, it became relatively easy for TikTok to get new partners.

5. Conclusion and Policy Implications

To summarize the key findings of the Chapter, based on the interviews conducted with tech companies, there are four types of incentives for tech firms to implement smart healthcare solutions (Chen 2022).

The first incentive is financial related. For instance, tech companies wish to improve their market share, promote brand value, venture into a new business sector, promote sales and look for new sources of profit. For governments, the incentive to implement smart healthcare solution is to control healthcare related cost.

Secondly, tech companies wish to improve their data accuracy and interoperability, and build algorithms to assist healthcare service providers to make decisions. Governments wish to improve healthcare efficiency by connecting different data sources and allocating limited financial resources efficiently by identifying the most urgent challenges.

The third incentive is associated with government policy. Tech companies need to answer government policy initiatives for Internet+Health, Healthy China 2030 Initiative, etc., and participate in smart city initiatives to build a high profile and maintain a good relationship with the government. Government contracts are rather lucrative and high-profile; winning government contracts suggests that the company is reliable and trustworthy with its solutions, thereby promoting the publicity of their relevant smart city solutions.

The last aspect is that of special incentives. Governments have incentives to create more employment and drive human resource development in AI. Furthermore, hospitals have the incentive to improve the healthcare quality for patients.

The DPE theory has laid down a foundation for analyzing the digital platform ecosystem in China. The growth of Alihealth and Xiaomi reinforces the DPE model. Alihealth had invested in several offline medication retailers in Gansu, Shandong, Hubei, Guizhou and Anhui to strengthen its online-to-offline (O2O) business model (Alibaba Health 2019). Both Alibaba and Xiaomi can be seen as multi ecosystems where they serve as platform leaders. The success of these two platforms depends on the quality of participants on the platform. Both the companies had proved their success in adapting to the changing market demand, in innovation and ecosystem orchestration.

Alibaba, the mother company of AliHealth, has expanded into the South East Asian market by investing and eventually acquiring Lazard; TikTok went into the Indian market and subsequently to North America and the European market; Huawei has developed its business in Africa, Europe and North America, establishing research centers across the globe. Moreover, Xiaomi opened its flagship shops all over Europe to promote its brand. Most of such expansions happen after a firm has established a dominant position in the domestic market.

Regarding the national level factors affecting Chinese firms going abroad, even though the domestic institution setting for doing business is improving, China still falls behind developed economies in terms of factors such as ease of access to loans. For instance, China ranks 31 in the ease to do business rankings (World Bank 2019). However, the ease of access to loans for private firms in China remains an issue, compared to state-owned enterprises.

On the industrial level, the Chinese government tries to promote national champions in different industries. This can be both advantageous and disadvantageous for Chinese firms to go abroad.

The One Belt One Road initiative has prompted Chinese firms to expand into the overseas market. Even though most firms which went abroad focus on infrastructure development in South East Asia, Africa and Latin America, more and more agricultural businesses are looking for opportunities overseas.

Tech firms such as Alibaba and Tencent, for instance, have invested in multiple start-ups, both domestic and overseas. To get into the Indonesian market for instance, Alibaba has invested in Lazada and Tokopedia to ensure full access to the market.

Yet, companies like Huawei have become victims of the nationalist internationalization strategy of China, thus losing contracts in all developed markets.

On the firm level, Chinese tech companies have certainly moved up the value chain and developed innovative business models and new strategies for expanding into the IoT market.

By market capitalization, Alibaba, Tencent and Ping An Insurance rank first, second and fourth respectively in China, with market values of $453 billion, $436 billion and $283 billion respectively (Business Insider 2019). Over the past 20 years or so, companies such as Alibaba, Xiaomi and Huawei have climbed up the value chain. These firms have experienced hurdles in their expansion and internationalization processes. They have benefited from the Chinese internationalization policy, yet suffered from the recent nationalist government internationalization strategy. These businesses have climbed up the value chain for sure and established different business models exploring the Internet of Things market. For further research on this subject, the author has provided a good example for studying Chinese tech companies and a new angle for looking at Chinese firms.

References

3M. (2020). 3M Science. Applied to Life. 3M United States. Retrieved March 14, 2020, from https://www.3m.com/.

Accenture. (2020). How we work with SAP. Retrieved March 14, 2020, from https://www.accenture.com/nl-en/services/intelligent-platforms-index.

ALIBABA HEALTH Information Technology Limited. (2019). ALIBABA HEALTH Information Technology Limited 2019 Annual Report (18). Retrieved from http://doc.irasia.com/listco/hk/alihealth/annual/2019/ar2019.pdf.

Alipay (2020-03-10) [Computer software]. (2020). Retrieved from https://mobile.alipay.com.

Banalieva, E. R., and Dhanaraj, C. (2019). Internalization theory for the digital economy. Journal of International Business Studies, 50(8): 1372–1387.

Baidu. (2020). Baidu Cloud Services. Retrieved March 14, 2020, from https://cloud.baidu.com/campaign/Annualceremony-2020/index.html.

Bloomberg. (2022, August 9). Tiktok owner ByteDance pays $1.5 B for China Hospital chain in Health Foray. Bloomberg.com. Retrieved November 8, 2022, from https://www.bloomberg.com/news/articles/2022-08-09/bytedance-pays-1-5-billion-for-hospital-chain-in-health-foray.

Bosch Global. (2020, March 12). Guided parking with parking sensors in the smart city. Retrieved March 14, 2020, from https://www.bosch-connectivity.com/.

Buckley, P. (2019). China goes global: provenance, projection, performance and policy. International Journal of Emerging Markets, 14(1): 6–23. https://doi.org/10.1108/ijoem-01-2017-0006.

Burning Rock. (n.d.). Burning Rock Medical. Retrieved March 18, 2020, from https://www.brbiotech.com/home/index.

Business Insider. (2019, July 26). These are the 14 biggest Chinese companies based on market cap. Retrieved from https://www.businessinsider.nl/14-biggest-chinese-companies-based-on-market-cap-2019-7/.

China Industrial Economic Database. (2018, June 1). China Export of ICT related products (in Chinese, translated by the author). Retrieved March 14, 2020, from https://www.ceicdata.com/zh-hans/indicator/china/exports-ict-goods.

Chen, N. (2022). Stakeholder power analysis of the facilitators and barriers for telehealth solution implementation in China: A qualitative study of individual users in Beijing and interviews with institutional stakeholders. JMIR Formative Research, 6(1). doi: 10.2196/19448.

Cesuroglu-de Langen, T. (2016). Integration of a personalized health care model into health systems and policies in Europe.

Coronavirus search trends. (2022). Google Trends. Available at: https://trends.google.com/trends/story/GB_cu_JSW_pHABAADqAM_en (Accessed: 05 July 2023).

Deng, P., Delios, A., and Peng, M. W. (2020). A geographic relational perspective on the internationalization of emerging market firms. Journal of International Business Studies, 51(1): 50–71.

Deloitte. (2020). Deloitte CN LSHC 2020 Global Health Care (in Chinese) [PDF document]. Retrieved from https://www2.deloitte.com/content/dam/Deloitte/cn/Documents/life-sciences-health-care/deloitte-cn-lshc-2020-global-health-care-zh-200211.pdf.

Dunning, J. H. (2001). The eclectic (OLI) paradigm of international production: Past, present and future. International Journal of the Economics of Business, 8(2): 173–190.

Dunning, J. H., and Lundan, S. M. (2008). Institutions and the OLI paradigm of the multinational enterprise. Asia Pacific Journal of Management, 25(4): 573–593.

Eisenhardt, K. M. (2012). Kathleen M. Eisenhardt on Case Study Research Methods. Edited by Li (Chinese Translation), P. P., Cao (Chinese Translation), F. Peking University Press.

European Commission. (2020, March 13). COVID-19: Commission sets out European coordinated response to counter the economic impact of the Coronavirus* (Press release). Retrieved from https://ec.europa.eu/commission/presscorner/detail/en/ip_20_459.

Financial Times. (2018, March 19). Alibaba invests another $2bn in south-east Asia’s Lazada. Retrieved March 14, 2020, from https://www.ft.com/content/44f3a1fe-2b44-11e8-9b4b-bc4b9f08f381.

Financial Times. (2019, December 6). Facial recognition: how China cornered the surveillance market. Retrieved March 14, 2020, from https://www.ft.com/content/6f1a8f48-1813-11ea-9ee4-11f260415385.

Financial Times. (2020, March 13). Huawei Technologies. Retrieved March 14, 2020, from https://www.ft.com/stream/254cd19f-4724-4c89-9230-926e8201a823.

Findlay, S. (2019). TikTok launches education push in India. Retrieved from: https://www.ft.com/content/a729232a-eff2-11e9-ad1e-4367d8281195. (Last Accessed: October 21st 2019).

Foxconn. (2020). Foxconn Profile. Retrieved March 14, 2020, from http://www.foxconn.com.cn/GroupProfile.html.

Grasselli, G., Pesenti, A., and Cecconi, M. (2020). Critical Care Utilization for the COVID-19 Outbreak in Lombardy, Italy. JAMA. https://doi.org/10.1001/jama.2020.4031.

GSM Association. (2019). Development with the Mobile China Economy. Retrieved from https://www.gsmaintelligence.com/research/?file=f3b217bd4616654f710854b0735b0c00&download.

Google Trends. (2022). COVID-19 Search Trends - United Kingdom. Retrieved from https://trends.google.com/trends/story/GB_cu_JSW_pHABAADqAM_en.

Haier. (2019). Haier Presentation at the National Conference Business Management Conference in China, Beijing, China.

Helfat, C. E., and Winter, S. G. (2011). Untangling dynamic and operational capabilities: Strategy for the (N) ever-changing world. Strategic Management Journal, 32(11): 1243–1250.

Helfat, C. E., and Raubitschek, R. S. (2018). Dynamic and integrative capabilities for profiting from innovation in digital platform-based ecosystems. Research Policy, 47(8): 1391–1399.

Hennart, J. F. (2019). Digitalized service multinationals and international business theory. Journal of International Business Studies, 50: 1388–1400.

He, S., Khan, Z., Lew, Y. K., and Fallon, G. (2019). Technological innovation as a source of Chinese multinationals' firm-specific advantages and internationalization. International Journal of Emerging Markets, 14(1): 115–133. https://doi.org/10.1108/ijoem-02-2017-0059.

Hull, D., and Zhang, C. (2019, October 23). Elon Musk opened Tesla's shanghai gigafactory in just 168 days. Bloomberg.com. https://www.bloomberg.com/news/articles/2019-10-23/elon-musk-opened-tesla-s-shanghai-gigafactory-in-just-168-days#xj4y7vzkg.

Huawei. (2018). Healthcare—Huawei solutions. Retrieved March 14, 2020, from https://e.huawei.com/uk/solutions/industries/healthcare.

Huawei. (2018). Huawei Cloud Service. Retrieved March 14, 2020, from https://hcs.huawei.com/portal/en-us/.

Huawei. (2018). Who we are. Retrieved March 15, 2020, from https://huawei.eu/who-we-are.

Huawei Vmall. (2018). Huawei Vmall Wearables . Retrieved March 15, 2020, from https://www.vmall.com/list-59.

IDC. (2022). Smartphone market share – market share, Available at: https://www.idc.com/promo/smartphone-market-share (Accessed: November 8, 2022).

IDC. (2022). Wearable Devices Market Share, Available at: https://www.idc.com/promo/wearablevendor/vendor (Accessed: November 8th 2022).

IDC. (March 5th 2019). IDC Reports Strong Growth in the Worldwide Wearables Market, Led by Holiday Shipments of Smartwatches, Wrist Bands, and Ear-Worn Devices. Retrieved from: www.idc.com/getdoc.jsp?containerId=prUS44901819. (Last accessed: October 21st 2019).

Infervision. (2020). Infervision. Retrieved March 18, 2020, from https://global.infervision.com/.

Intel. (2020, March 14). Intel® CoreTM Processor Family. Retrieved March 14, 2020, from https://www.intel.com/content/www/us/en/products/processors/core.html.

INVEOX. (2019, August 9). inveox - rethinking pathology. Retrieved March 18, 2020, from https://inveox.com/about-us/.

John Hopkins University. (2020, March 18). Coronavirus COVID-19 Global Cases by the Center for Systems Science and Engineering (CSSE) at Johns Hopkins University. Retrieved March 18, 2020, from https://coronavirus.jhu.edu/map.html.

Keep. (2020). About us, Keep. Retrieved March 18, 2020, from https://www.gotokeep.com/about.

Knoerich, J. (2019). Re-orienting the paradigm: path dependence in FDI theory and the emerging multinationals. International Journal of Emerging Markets, 14(1): 51–69. https://doi.org/10.1108/ijoem-04-2017-0123.

Linkdoc. (2020). Smart Assistant Diagnosis System. Retrieved March 18, 2020, from https://www.linkdoc.com/product/AI.

Levine, A. S. (2022, August 26). Tiktok, hospitals and tutoring apps: The many tentacles of chinese tech giant bytedance. Forbes. Retrieved November 8, 2022, from https://www.forbes.com/sites/alexandralevine/2022/08/24/tiktok-parent-bytedance-companies/?sh=3b50b9c6d236.

Levine, A. S. (2022). Tiktok, hospitals and tutoring apps: The many tentacles of chinese tech giant bytedance, Forbes. Available at: https://www.forbes.com/sites/alexandralevine/2022/08/24/tiktok-parent-bytedance-companies/ (Accessed: 05 July 2023).

Liu, J. (2019). The roles of emerging multinational companies' technology-driven FDIs in their learning processes for innovation. International Journal of Emerging Markets, 14(1): 91–114. https://doi.org/10.1108/ijoem-07-2017-0232.

Mobvoi. (2020). Mobvoi.com - TicWatch smartwatch TicPods Free . Retrieved March 18, 2020, from https://www.mobvoi.com/nl/types/wearable.

McGee, P. (2019). As iPhone 11 looms, Apple looks beyond the smartphone. Retrieved from: https://www.ft.com/content/789b745c-d096-11e9-99a4-b5ded7a7fe3f. (Last Accessed: October 21st 2019).

Nambisan, S., Zahra, S. A., and Luo, Y. (2019). Global platforms and ecosystems: Implications for international business theories. Journal of International Business Studies, 50(9): 1464–1486.

NASA. (2020). Airborne Nitrogen Dioxide plummets over China. NASA. https://earthobservatory.nasa.gov/images/146362/airborne-nitrogen-dioxide-plummets-over-china.

NXP. (2020). Processors and Microcontrollers, NXP. Retrieved March 14, 2020, from https://www.nxp.com/products/processors-and-microcontrollers:MICROCONTROLLERS-AND-PROCESSORS.

Pan, Z., Nguyen, H. L., Abu-Gellban, H., and Zhang, Y. (2020, December). Google trends analysis of covid-19 pandemic. In 2020 IEEE International Conference on Big Data (Big Data) (pp. 3438–3446). IEEE.

Parkinson, J., Bariyo, N., and Chin, J. (2019). Huawei Technicians helped African governments spy on political opponents, The Wall Street Journal. Available at: https://www.wsj.com/articles/huawei-technicians-helped-african-governments-spy-on-political-opponents-11565793017 (Accessed: 17 June 2023).

Qualcomm. (2020, February 25). Flagship Qualcomm Snapdragon 865 5G Mobile Platform Powers First Wave of 2020 5G Smartphones. Retrieved March 14, 2020, from https://www.qualcomm.com/news/releases/2020/02/25/flagship-qualcomm-snapdragon-865-5g-mobile-platform-powers-first-wave-2020.

Remuzzi, A., and Remuzzi, G. (2020). COVID-19 and Italy: what next? The Lancet. https://doi.org/10.1016/s0140-6736(20)30627-9.

Ruehl, M. (2019, October 8). China's leading AI start-ups hit by US blacklisting. Retrieved March 14, 2020, from https://www.ft.com/content/663ab29c-e9bd-11e9-85f4-d00e5018f061.

Sensetime. (2020). Medical Image Analysis-Core Technologies-SenseTime. Retrieved March 18, 2020, from https://www.sensetime.com/en/Technology/medical.html.

Sharma, D. D., and Blomstermo, A. (2003). The internationalization process of born globals: a network view. International Business Review, 12(6): 739–753.

Teece, D. J. (2007). Explicating dynamic capabilities: The nature and micro-foundations of (sustainable) enterprise performance. Strategic Management Journal, 28(13): 1319–1350.

Teece, D. J. (2018a). Business models and dynamic capabilities. Long Range Planning, 51(1): 40–49.

Teece, D. J. (2018b). Profiting from innovation in the digital economy: Standards, complementary assets, and business models in the wireless world. Research Policy, 47: 1367–1387.

The World Bank. (2022). COVID-19 Google Trends Dashboard. Retrieved from https://datanalytics.worldbank.org/covid_gtrends/.

UNCTAD Statistics. (2020, March 14). UNCTAD . Retrieved March 14, 2020, from https://unctadstat.unctad.org/wds/?aspxerrorpath=/wds/ReportFolders/reportFolders.aspx.

Wearable Technology. (2019, October 23). Wearable Technologies Booklet, 2019 MEDICA, Dusseldorf, Germany.

Waltze, S. (2019). Strategy Analytics: Apple Watch Captures Half of 18 Million Global Smartwatch Shipments in Q4 2018. Retrieved from: https://www.businesswire.com/news/home/20190227006074/en/Strategy-Analytics-Apple-Watch-Captures-18-Million. (Last Accessed: October 21st 2019).

Wearable Technology. (2019, October 23). Wearable Technologies Show 2019 MEDICA.

Wegene. (2020). WeGene - Personal Genetic Testing Platform. Retrieved March 18, 2020, from https://www.wegene.com/en/.

World Bank. (2019). Mobile penetration. World Bank DataBank. Retrieved from https://databank.worldbank.org/Mobile-penetration-/id/5494af8e.

World Health Organization. (2022). WHO Coronavirus (COVID-19) Dashboard. Retrieved from https://covid19.who.int/.

Xiaomi. (2020). Xiaomi Privacy Policy. Retrieved March 14, 2020, from https://www.mi.com/global/about/privacy.

Xinhua Newsnet. (2019). Huawei has invested hundreds of billions in R&D (in Chinese). Retrieved March 15, 2020, from http://www.xinhuanet.com/tech/2019-03/30/c_1124303527.htm.

Yitu Technology. (2020). Smart Health Yitu Technology. Retrieved March 18, 2020, from https://www.yitutech.com/cn/business/yitu-heathcare.

Yin, R. K. (2017). Case study research and applications: Design and methods. Sage publications.

Appendix

Interview Records with VP of Wearables and Health Unit of Huawei

Part I: Personal Health related products and smart home products

1. What are the products offered by Huawei for sports coaching, health monitoring, and health risk prevention? Do Huawei products offers continuous monitoring and chronic disease management functions, particularly for the aging population? Is there any design for the elderly to live independently at home, particularly for fall detection?

The main business model of wearables in Huawei is focused on hardware. The products targeting the elderly is in design, with the main design philosophy targeting at user demand. The plan is to transfer some of the functions of the children's watch to the devices designed for elderly, for instance, the GPS location function. Possible other functions to add to the design include fall detection automatic alarm, heart rate monitoring, etc. There is a smart care plan designed for the elderly.

Huawei will increase interaction with elderly, and try to take in suggestions from the elderly users, and co-develop with users targeting at their needs.

2. In the previous stage of study, I read about delve into the value of he wearables. It plans to do so by improving life quality, enhancing environment sensibility, promoting user experience. Huawei has proposed establishing the digital twin of users, and to realize digitize the environment of human beings, and to improve the interactions with the environment of users, and to warn patients of healthcare risks. Are these concepts which has been used in product design?

Normally we use phones, pads, and sports watch, the Huawei Sports app, Hlink App and router to interact with users. Huawei Hi Link App has been designed for smart application scenarios in home entertainment, home lighting, smart health, energy management, home security, automation, routers, smart TV boxes, lighting, environment, interior design, all coverage routers, Wifi, kitchen, audio, smart sleep monitoring, etc. There are about 50 core partners, and in the future, it is possible to design a comprehensive platform, for instance to add humidity and temperature sensors into smartphone and smart watch design.

Huawei has promoted the use of interaction depending on the maturity of the technology. Now Huawei is focusing on delving into the health related function for wearables and terminals (telephones). For instance, there are temperature sensors on phones, and in the future can detect temperature inside and outside home. The same design is planned on smart watches. Meanwhile, there are temperature and humidity sensors on smart home devices.

3. **Now the wearables can perform sports coaching, smart sleep monitoring, smart sleep, stress management, BMI management, skin management. What are the functions used most by users? What are the feedbacks?**

The most used functions are sports functions. Now sports management, sleep, stress monitoring, skin management can be achieved with Huawei Phones, other products with Huawei brand or under the Huawei ecosystem. Now it is possible to detect characteristics of skins, and to differentiate between red areas and brown spots. The devices distinguish between pore, blackhead, lines, color spots and red areas. Afterwards, it will grade the skin quality of users, and offer skin care advices for users.

Now the data about sports, and sleep related functions can be processed on device chips with edge computing to update the algorithms. The data is protected on devices, chips and in cloud. Only photos and insensitive data are uploaded to the cloud. With finger print, weight, sports, sleep, and heart rates and other sensitive data is processed on the device.

The sleep monitoring band (Reston) was developed with third-party partners, with smart weight scale developed independently by Huawei.

The edge computing advantage is in its flexibility and security regarding privacy. The 5G network is helpful for developing edge computing. With Huawei's own chips, and algorithms, it is helpful to perform edge computing.

4. **Do Huawei and other ecosystem businesses offer systematic health risk monitoring and health life advices, and to predict health risk for users?**

The monitoring center is based on phones. In the meantime, there is Huawei sports app which connect Huawei devices and partner devices. Now there is no comprehensive platform to monitor health risks for users. The main problems are among all the methods to perform blood pressure monitoring, the CFDA only approves blood pumping. Monitoring blood pressure 24/7 over blood pumping will disturb users particularly their sleep. It is not possible to commercialize the blood pressure monitoring function on Huawei watches and bands.

5. **Has Huawei formed the IoT device ecosystem? Has Huawei formed an IoT device platform controlling all devices? What is the control terminal?**

With the Huawei smartphone as the IoT ecosystem devices, now there are a few application scenarios: smart office/ smart gaming, connected cars, smart health and smart home. The control terminal is the smart health app on the phone. Home smart devices can be controlled via smart audio devices (Little Yi). Huawei routers can keep the devices online with the Hi Link app connect devices.

6. **What are the smart home and smart health (nutrition analysis, smart audio, sleep monitor, smart scale, smart mask) devices offered by Huawei? What is the control terminal for Huawei smart home devices?**

The advantage of Huawei is about its terminal devices and the strength in telecommunication (I assume they are referring to 5G). There are devices which the Huawei can involve, but will offer interfaces for connection. Now the nutrition analysis function can be realized via Huawei Mate 20 smart food detection. It is possible to use camera and distance detection to identify how many calories are contained in an apple. The algorithms can automatically tell the size of the object and the amount of calories contained.

For smart home devices, Huawei is planning on commercialize smart TV, and hope to include all smart health functions on smartphones.

7. **Huawei is planning on offer personal health records services for users? If there are such plans, how to realize real time EHR system update?**

Huawei offers the solution packages for user to realize healthy living and disease prevention. The function is realized via Huawei hardware terminals to monitor heart rate, sleep, sports, and via connecting third party devices to monitor blood pressure and blood sugar level. In the meantime, Huawei hopes to help users to perform chronic disease management, and outpatient long-term rehabilitation plans for users. Huawei also hopes to offer telemedicine for patients who need long-distance services. In the meantime, there will be issues because currently doctors only rely on patients to describe their conditions, and the descriptions performed by patients can be objective and inaccurate.

Huawei is carrying out experiments for application of wearables in medical scenarios with Beijing 301 hospital (military hospital) in the area of cardiovascular health research. In the future, Huawei hopes to perform preventive healthcare, to lower the burden for hospitals and patients. Now the function has already been commercialized on smart watches. The AI automatic warning for irregular heart rates can only be commercialized after the CFDA certification.

Now there is no user personal health records functions. It is because user health data is sensitive in the sense of privacy. In the meantime, the industry of wearables have not set up technology standards. Now the CFDA use the same medical device technological standards to determine whether wearables can be used in medical scenarios. It takes about 5–10 years to get the CFDA approval; this long waiting time had resulted in the lack of commercialization for certain functions. For instance, the two-point ECG monitoring cannot be commercialized. In many practices, for instance, the blood pressure monitoring requires continuous monitoring; but there are not relevant technical standards for wearables to perform health monitoring. The only legal method of monitoring is to use blood pumping for the monitoring. The 24-hour continuous monitoring will disturb the sleep of users. Normal users do not necessarily need such precise data which comes with blood pumping . Now there is no such technical standards. It is a loss for users, and for the doctors.

To thoroughly evaluate users' health risks, it is necessary to evaluate blood oxygen, heart rate, and life style, etc. Such system will be available with the efforts from Huawei and other institutions. Huawei is telecommunication device manufacturer, and cannot independently finish the EHR system construction. The partnership with individual hospitals was formed on an individual basis with hospitals. Without a uniform EHR system on a national level, electrification of medical resources and interoperability of data between different hospitals, establishing EHR for users does not have much value.

8. **Does Huawei take use of data collected from users to train algorithms to offer personal health risk monitoring? Are these algorithms running on devices or on cloud platforms?**

The data and algorithms trained in labs of Huawei may not fit to individual users. Therefore, by update algorithm with data collected from users, the algorithms become more precise. Before the algorithms become available to users, there will be 1–2 years of clinical trial programs at institutions such as Harvard Medical School. After obtaining the missing data samples from users, algorithms will be updated or updated with certain data subjects with diseases.

Huawei has carried out data processing following GDPR standards in China and in European markets after GDPR was carried out.

Health data will only make sense when continuous monitoring is performed, in the meantime, it is necessary to conduct correlation analysis for health related data. Huawei as a tech company alone cannot perform such tasks. There is a need to form partnership with more institutions for instance, with governments to share health related data on a regional and national level. For instance, blood sugar, blood pressure needs to be monitored on a daily basis at the same time. Before establishing the national level EHR system, it impossible to perform population health management. Therefore it is necessary to have national and regional level EHR system.

9. **Can Huawei's products connect with third party platforms? Will Huawei share data with third party partners? or share analytical results? Does Huawei share data with insurance companies, clinics, hospitals, hospitals, governments and other institutions? Is there any related business model about data sharing? What is the business model to partner with Omron, Yu Yue, Biolight, Dnuse and Johnson and Johnson?**

It is sensitive to share data with third party manufacturer. Even with users consents, and sharing health related data can be legal, it is impossible to ensure third party partners adhere to laws and regulations. Therefore, Huawei is exploring business model based on data.

Huawei is working with third party partners, such as Omron and Dnurses to track blood pressure and blood sugar level changes. Huawei mainly use its terminals to keep track of the biometic indicators.

Huawei smartphones is pre-installed with Wedoctor, DXY, Chun Yu Doctor apps to make it convenient for users and give users better healthcare experience and to offer one-stop health services.

Now the South Africa insurance company—Discovery is working with Huawei to offer free wearable devices, which charges users on a term basis. When users reach a certain number of steps daily, the monthly term payment can be exempted. Users can lower their insurance premium by using the device long term. In the meantime, it is possible to lower health risks for users and form a positive feedback system.

Now the partnership with medical institutions is mainly for research purposes.

10. **Has Huawei established a data centered business model for wearable devices?**

There is unlimited value embedded in wearables. Now the value detected has been only a tip of the iceberg. In the future, possibly everyone will have 1–2 wearable devices. Therefore it is valuable to mine the value from the data collected from wearables.

Now there is no such business models, because there is no clear legal definition regarding the ownership of the data. Huawei has listed establishing a data based business model in the business development plan. Finding a data based business model has gained consensus for the IoHT industry.

11. **How to protect health related data of users? After the launch of GDPR, did Huawei take additional measures to protect users health relate data? Did GDPR affect the sale of Huawei wearable devices in Europe?**

Wearable devices collect and store data. After the launch of GDPR, processing and storing data has become slower and more intricate. The **AF detection and blood pressure monitoring** will possibly not be commercialized in Europe after being developed. Now the function most used in Europe is sports related function. Now the wearable devices offered in domestic (Chinese) and overseas market are almost the same. The overseas devices do not come with apps preinstalled such as Wedoctor, DXY, and Chun Yu doctor. In the future, there might be difference. In the European market, Huawei devices connected with less third-party devices, and mainly performs health data tracking (writing data down, but no coaching) functions.

12. **How much of the overseas market sale account for the total sale of Huawei wearable devices?**

When Huawei develops its overseas market, the development is mainly for domestic consumers. The path usually taken is to learn from the advanced concepts and to develop new functions based on the users needs in the domestic market (the Chinese market). Then Huawei tries to promote the products in the overseas market.

Wearable devices of Huawei are sold both on the domestic and overseas market. The functions are not the same. Wearables sold in the overseas market can lose some of the functions to connect with third party products. For instance, the smart watch sold in the EU market only has the basic function of health tracking, but not coaching.

What are the advantages and disadvantage factors to develop the overseas market?

In Europe and the US, wearable devices manufactured by Chinese companies face strict regulations. Meanwhile to develop health related functions, it is necessary to consider the local culture. Therefore, it is not possible to delve in the value for wearables compared with the Chinese market. The development process is oriented at users in the Chinese market. There are also price and function related concerns.

Huawei adhere to the local regulations strictly, and therefore it is not possible to restraint the sale of Huawei products legally (seems to contradict the trade war).

There are few restrictions for the sleep and sports related function in the overseas market. The functions are not going into the medical directions yet. Whether in the future there will be restraints towards wearbles and medical related functions embedded is yet to see.

What are the role of overseas development center for Huawei?

The wearables are mainly developed in the domestic market. The Huawei HQ in China is a coordination center to manage all the businesses for Huawei overseas growth. The location choice for the overseas development center was chosen based on technology, for instance, like the algorithm development center in Finland, and the hardware development center in Germany, and the design center in France (for fashion), etc. The sample in the labs for developing algorithms is relatively small. It is mainly developed in labs and does not use data from users, and will not transfer data from China to overseas labs or transfer data from overseas to China. It is possible to update the algorithms later at the clinical trial or commercialization stage with larger size of data.

13. **What are the roles played by wearables in the value based healthcare system?**

The center of establishing a value based healthcare system in based on data. Without data, it is difficult to perform preventive healthcare. Therefore, smart wearables plays an important role. Data needs to be precise and cover a wide range of categories to realize extracting data from wearables.

Now the commercialization of wearables in the hospital use scenario is limited.

The ideal condition is there is a national wide health data sharing platform, where all doctors and hospitals can see the data. In the meantime, there is a unified EHR system on a national basis. In this way, it is possible to use wearables on a precise, wide and useful basis. With no technical standards for wearable use in the healthcare industry, doctors would shake their heads (and use the wearables) when they see devices without CFDA approval.

It is not realistic to ask the basic medical insurance schemes in China to cover the cost of wearables. If they try to cover the cost, the total amount of expenses may grow.

Huawei is just a device manufacturer. Our goal is to lower the price of our devices. while offering the best quality as we can to consumers.

The China National Development and Reform Commission is running a trial program in Qing Dao to test 5G city and require the establishment of telemedicine user case scenarios.

Xiaomi Interview Records with investor director of MIUI Business Unit Personal Users, Wearables and smart home devices

1. **What are the continuous health monitoring products for personal users?**

   Smart blood pressure monitor, smart thermometer, smart humidity monitor which can turn on and off automatically for home humidifier, or air conditioner), and smart scales are produced by Xiaomi ecosystem enterprises.

   Wearables ecosystem is made up of software and hardware, software is usually developed by Xiaomi ecosystem enterprises, with the data collection made on the software end. The software do not belong to Xiaomi, and all the data are connected by software developed by Xiaomi ecosystem enterprises. Xiaomi watch and bands are manufactured by Huami. As far as I know, Xiaomi has not obtained the data from the ecosystem enterprises and performed user profile analysis.

   All the wearable software services are in the integrated in the Mi Home platform. Xiaomi has not developed a business model based on data. Now the data collected from Xiaomi band is controlled by Huami. The company may perform user profile analysis with the data collected. There has not been a business model which allows Xiaomi to recommend weight control or sports related service packages to users for overweight users.

   Xiaomi now has a business model based on hardware sale. The service based business model has not been promoted. Take the Xiaomi watch for example, Huami has used a double brand strategy, where they try to promote both Amazfit and Xiaomi Watch as the brands for the Amazing Fit watch. When Xiaomi decided to invest in Huami, there was no consideration about delving into the data centered business model on the backstage. The business model is to sell hardware (except smartphones and television boxes) to attract users, and to nurture the user habits, and to attract users to the Xiaomi platform. In the early stage of internet based business, the business model is about get users to use a certain product, and to nurture user habits (for instance, Tencent acquired a large amount of users via platforms such as QQ).

The band's main market is focused on the domestic market, and some are also sold in the overseas market.

All the IoT products belong to Mi Home brand. There are over 20.3 million active users per month. There are over 60–70 terminals offered by Xiaomi and over 10 active platforms has more than 1 million daily active users. There are two platforms covering IoT businesses, like Mi Home and Mi Sports. Xiao Ai (Like Siri, is the voice assistant for Xiaomi ecosystem products) is an extension, with Xiao Ai able to control smart cleaning robots. Xiao Ai can learn about the language habits of users and adjust its communication style. Xiao Ai can also connect with the voice to text translation software, but it will not be done via Xiao Ai the extension itself.

2. **Xiaomi has proposed the business plan to improve interaction with users, with the envionrment and to improve life quality of users. Currently Xiaomi products cover a wide category of smart wearables and smart home devices, such as smart scale, smart toothbrush, smart watches, bands, smart lighting system, smart air purifier, smart rice cooker, etc. In the healthcare industry, Xiaomi has invested in Andon Health, and now in the Xiaomi You Pick Online Mall, Andon Hipee Pen, Xiaomi iHealth blood pressure monitor, smart thermometer, smart Gluco monitor are now available for consumers. Does Xiaomi and the ecosystem partners plan to offer systematic health monitoring and smart health advice for users? How to implement the development strategy whereas Xiaomi use wearables and smart devices to enter the telemedicine and big data in health market, and build the health ecosystem with users?**

When Xiaomi make the investments in the field of IoT, it does not include a strategy to focus on healthcare. In the early stage of IoT devices, there is no direction because it is a blank market with a wide category of choices. Xiaomi does not choose a specific industry or the business to invest in, but rather focus on user needs. Xiaomi invested based on the 80%/80% rule, where the product design focuses on meeting the needs of 80% of users and 80% of their needs. It tries to face the mass market and offer standardized choices with a large user base for general use. Xiaomi also tries to focus on the core functions of the product and to optimize the function to save R&D costs (Rice cooker, air purifier, air conditioner, etc).

The internet based business and IoT based business models differ in that Baidu, Alibaba and Tencent (BAT) has their own main business focus. For instance, Tencent focused on social media, gaming and new media; Alibaba is running B2B and B2C platform, and focus on transactions. Xiaomi does not have a major strategy in mind when it made investment for IoT based products. It invested in a wide range of areas, and focus on the area of business with the biggest growth potential. The business unit with the best growth rate receives the most support.

In the early stage of product development, there is a lot of space for growth and it is barbaric style of growth. The investment to develop smart shoes (shoes with chips inside) was made under such business strategy. Now Xiaomi is not just focusing developing smartphones or a specific product; instead Xiaomi is trying to meet all daily needs for Mi fans. In the end, Xiaomi hopes to reach the goal the meet all the daily needs for Mi fans. The trial program of producing smart shoes runs under the strategy that Xiaomi will form a product a matrix for our users.

3. **Does Xiaomi has any plans to provide data from wearables for users to have personal health records?**

Synchronizing data from Xiaomi band takes Xiaomi Health app to do so. Now the app is controlled by Huami. Xiaomi is considering take the app back from Huami and develop it. Now the health related business is not on the Xiaomi priority list. After Xiaomi went public, we are now focusing on businesses which will benefit our balance sheets, such as promoting the sale of online ads. After the listing of Xiaomi stocks, Xiaomi puts a lot of emphasis on KPI.

4. **How does Xiaomi handle data from wearables? Does Xiaomi use algorithms developed by Xiaomi to analyze the data? Are these algorithms running on the cloud or on the device? Will the collected data used on training algorithms to offer personal health risk monitoring?**

We now acquire data from app ending to analyze our user profile; our partnership with them adds a condition that if the app tries to read users' contact list, then we turn it down. There is no data based business model yet. In the future, the healthcare industry, the elderly care industry and the education industry are going to drive economic growth in China. Therefore Xiaomi will probably invest in these industries in the future.

In the future, there will be a business model focusing on data sharing. Xiaomi might explore the partnership with insurance companies to perform customer segmentation and to share data. The band is mainly on sale in China.

5. **Does Xiaomi share data with third party partners or share the analytical result?**

Now it is possible to share data, with the sharing channels rather simple. For instance, to share the data on the Xiaomi band on a weekly and monthly basis and send it via pictures. There is no way to share the data automatically in the Mi Fit app. In the app, only users can initiate data sharing via simple ways such as send pictures over wechat . Xiaomi has not shared data with third parties such as gyms. In the future, Xiaomi may share data with gyms and doctors.

The main business model for Xiaomi IoT is about selling the product itself, instead of selling services. To attract the users, to develop a product based business model, sell the problems via new retail channels, and internet-based businesses are three main drivers' for Xiaomi's growth. The internet based business mainly comes from advertisement; among the 16 billion internet based businesses, about 10–12 billion arise from advertisement related sales. In the app store, it is also divert users from app store (to get users to download an app). The other methods to divert users to share information via browsers, or apps such as Tou Tiao. The Xiaomi ecosystem is supported by the MIUI platform, with the internet based businesses come mainly from service fees, such as listening to songs, video gaming, and streaming videos.

Data is valuable, for instance, to share pictures via taking photos and sharing it via plastic surgery hospitals. Now there is no data based business model.

From the Xiaomi annual report, now the sale for IoT and consumer products is around 43.8 billion RMB. Xiaomi products has reached the sale of around 8.4 million sales globally, with the Mi Band ranks second on the global wearables sales volume. Now the Mi Home platform connects around 150.9 million devices. Third-party devices, such as Ikea lighting devices, has been able to connect to Xiaomi Mi Home platform, and controlled via Xiao Ai. Now there are about 203 million active users on Mi Home with over 50% of users come from non-Xiaomi smartphones.

For users there is privacy related concern on whether data has uploaded to private cloud or public cloud.

6. **How does Xiaomi deal with the privacy related risks of users' health risks? Did the launch of GDPR in May 2018 affect the sale of Xiaomi wearable products and other health related products? Did Xiaomi take additional measures to protect users' personal health data?**

Making an advertisement promotion demands conversion rates. From the advertiser's perspective, targeting users as accurately as possible means a higher conversion rate, and better effects of the advertisement.

Our legal department will consider whether the specific type of third party services Xiaomi violates privacy protection rule.

After the launch of GDPR, when Xiaomi bought data from quest to analyze the usability of apps, the contract takes 3 months to go through. There is no records to show how the data was collected. In the end, the department analyzing the data has to ensure the department will take all the responsibility for violating data protection law. Domestic Chinese companies are more and more responsible for protecting data. The advertisement is mainly about user profile, and targeting users by name, geography, age. There are many internet focused businesses in China expanding their markets overseas, which makes the firms running the business focusing more and more on privacy. When new users register, they need to be informed of how data will be collected, used and the data flows.

Xiaomi smart home products and bands are targeting domestic markets only, with domestic and overseas users (China and overseas) makes an 1:1 ratio, with about 300 million users in total.

Now Xiaomi and Zhong An Insurance (an internet based insurance company) develops the steps based insurance, with the insurance cuts premiums if the user can reach the steps target per day (measured by their smartphone, wearables, etc.). Users can be diverted from Xiaomi platforms to Zhong An (now the cooperation seems to have stopped). The data collected by blood pressure monitor, scale, and bands will be insurance related business in the future.

7. **How does Xiaomi cooperate with Microsoft or Signify to develop a new business model for IoT? Will Xiaomi share data collected from IoT devices with Microsoft? Is Xiaomi mainly responsible for hardware development?**

Data will be of use in the future to develop new business opportunities. Apart from Signify lighting products, Mi Home can also connect to Ikea and other hardware. Our strategy is to build a Mi Home platform first, and to use it as a control terminal for users inflow and the smart ecosystem construction. The data about timing of lights on and off is valuable, but the use it is not yet known. The smart health related data is not that useful for now, because there is no insurance business within Xiaomi. In the future, with such businesses, it will be useful.

8. **Xiaomi has promoted Xiao ai as the AI audio assistant (an equivalent of Siri or Google Assistant). Now what are the wearables and smart home devices controllable via smart audio assistant system? Is it possible to record user sleep, pregnancy weight, food and nutrition analysis? What is the current business model for smart wearables and smart home devices of Xiaomi?**

Xiaomi is building an ecosystem with the control terminal as the Mi Home app.

Xiao Ai is another interaction method. When Xiaomi starts to experiment with smart home devices, we tend to believe routers can serve as the control center. Before Mi Home, the Mi Wifi App serves the same function. By December 31st 2018, Xiao Ai has reached a monthly active user of about 388 million.

When Xiaomi started to sell the smart rice cooker, Xiaomi intends to sell rice, nutritional breakfast, and packaged food, and other related products like the rice cooker. Now it is possible to control cooking and warming functions of the rice cooker. Currently the nutrition analysis, and cooking guidance function will be able to commercialize in the next 3–5 years.

For instance, to realize the nutrition analysis function, smart fridges can be a better use case scenario. The cameras inside the fridge or RFID scanner inside the fridge can recommend menus based on personal preferences and remind users to order missing materials based on the dish choices online with fresh vegetable delivery services. It is possible to remind users about the food expiration dates. For instance, there are eggplants and peppers, the fridge will recommend the users to cook Di San Xian (Fried Potato, Eggplant and Pepper in Garlic Sauce). If the users do not have the potatoes, users can order immediately with the fridge.

9. **What are the policy implications for initiatives such as "Health China 2030" for Xiaomi to develop its smart health industry?**

Macroeconomic policy carried out by the Chinese government, which can facilitate the smart health industry development does play a role in Xiaomi's investment decisions. Before making an investment, Xiaomi need to narrow down the market size for the industry, check whether the specific project and the specific team fit into the picture. When Xiaomi invest in IoT ecosystem products, Xiaomi follows a bamboo forest roadmap to invest in a wide category of companies and businesses, to counteract the business cycle effects for tech companies. Xiaomi can only succeed with the ecosystem functions like a bamboo forest, with a multiplication of new shoots to create an ecosystem.[6] In this way, Xiaomi only focused on developing important products smartphones, tablets, TV and routers.

10. **Xiaomi has invested in Ping An Good Doctor (offers patient registration and online medication sale services), DXY.com (Internet based hospital which offers clinic management, online doctor consultation, online forums for doctors and offline clinics, etc.),**

Jeejan.com (elderly care facilities, outpatient management, etc.), Ibabycenter (pregnancy prepation online and offline), Fang Cun Doctor (professional outpatient management), Tough Workout.com, 8HSleep.com (smart mattress), Miaomiaoce.com (Genetic Testing, Cancer Prevention), Soocare.com (smart toothbrush), and other ecosystem companies in and outside the Xiaomi ecosystem. Does Xiaomi has plan to cover disease prevention, diagnosis, outpatient management, female health (pre-pregnancy and post-pregnancy), and the entire healthcare continuum?

Xiaomi is not put the focus of investment on the healthcare industry. Xiaomi has invested in several areas, with an attempt to form a product matrix. Whether it is healthcare, transportation, housing, retail user scenarios, Xiaomi hopes to be part of it. When the IoT industry just started, Xiaomi takes a barbaric growth strategy. Now Xiaomi has a strategy (when making investment for IoT related business). Hopefully in the next 3–5 years or 5–10 years, making an investment decision depends on the industry, the competition track, the opportunities embedded and the specific products.

Xiaomi focuses the sports related functions for the wearables and the most important one is to interact with users. Within the 200–300 companies Xiaomi invested in, there are around 20–30 firms focusing on healthy living. Xiaomi has invested in around 150–180 ecosystem companies focusing on IoT related businesses. There is not an investment strategy in there. For instance, when Xiaomi invested in toothbrushes, Xiaomi is not investing in health related segment. The logic embedded in investment was not that clear in the beginning. The investment on smart toothbrush focuses on non-durable consumer goods, instead of smart health industry. The investment logic is to invest in consumer products that can interact with users, which consumers use with high-frequency, with a wide product interest.

Xiaomi Interview Notes with Business Manager of Wearable Business Unit—Wearable Device Data Collection and User Experience

Part I: Xiaomi smart watch, band, user experience and major markets

1. Currently, what is the proportion of users of Xiaomi smart watches and bands in the domestic and overseas market?

   The business scope of Xiaomi for wearables covers Mi Band and Mi Scales.

Will Xiaomi tries to bundle the data from the watch, band and scale and build user health profile?

In the long term, data from people's watch, band, weight scale are biometric data which falls under the personal health umbrella theoretically. All the devices falls into the big health umbrella. Taking use of data from multiple devices in the near future would be difficult, which is related to the overall industrial development and practice. Many of our competitors, for instance, Apple, only provide wearables. Only the Xiaomi ecosystem covers many categories of IoT products and we are the only one following such unique business model. Covering as many IoT product category as possible is also the Xiaomi IoT ecosystem, which we are developing. Creating the IoT device network and the link between different devices is not easy, because it involves coordinating very complicated relationships. The forming of the network involves company management and stakeholder interest; therefore, it is not easy to get synergy value from the IoT ecosystem.

The market has not reached the stage of taking use of the network value of IoT ecosystem. Taking use of the under-realized value of the IoT network will happen in the next 3–5 years. Combining various data sources into a single health initiative will be possible, and the market has not reached this stage. In fact, the data can only be useful with the amount of data and the accuracy of the algorithms reaching the threshold. These thresholds have not been reached yet. Xiaomi has not fully explored the commercial value of data.

Huami and Xiaomi are sharing data collected from wearables; the data sharing relationship has been set from the beginning since Xiaomi invested in Huami to become the sore hardware manufacturer. The ownership of the data is not related the manufacturer of the device. All data belongs to Xiaomi theoretically speaking. Xiaomi has invested in Huami. With both companies listed publically, it is not possible to discuss the issues in detail.

Does the difference in data protection regulation play a role in the sale of Xiaomi wearables in the overseas and local market in China?

The difference does has an impact on our sales; Xiaomi has upgraded our front-end privacy settings and do not need to worry about violating privacy regulations anymore.

Xiaomi wearables category does not include watches, only bands. Amazingfit is another brand created by Huami, and it has nothing to do with the Xiaomi brand. Huami tries to promote Amazingfit watches as Xiaomi smart watch, and wish to impress consumers with the Xiaomi brand. Xiaomi does not deny it.

Now the overseas market and the domestic market accounts for half and half of Xiaomi wearable sales. The sales of Xiaomi wearables grows at a faster rate in the overseas market than in the domestic market. The Xiaomi products overseas sales is relatively good in the European and the India market. There is a small proportion of sales in the Africa market, with the other markets accounting for the rest of Xiaomi wearable overseas sales. Before the annual report goes public, Xiaomi cannot disclose specific sales data (from the annual sales report, Xiaomi wearable domestic sales accounting for 61% of the overall sales till Q3 2018, compared to 81% from the previous year).

2. **Who are the main user groups for Xiaomi wearables in China? Who are the main users in the overseas market?**

Previously Xiaomi mainly cover users between 16–30 years old, and when it comes to specific products, target user group can expand (for instance, Mi Bunny, etc.). The normal user group is between 10–40 years. Except for children and the elderly, other population subsections all belong to Xiaomi target user group. There is no big difference between target users and the actual user groups.

Will brand positioning affect Xiaomi user targeting in overseas and domestic market?

Xiaomi brand stands on the same level as Samsung, with different brand perceptions will affect users' buying decision. In the indian market, it is possible the rich Indians will consider buying Xiaomi. In China, rich Chinese customers will not consider Xiaomi products. In the domestic market, Xiaomi is a commoner's brand. It is like MUJI product positioning in Japan and in China.

When it comes to customer segmentation, Xiaomi would consider user age, gender, and income. Normally Xiaomi would not consider such details in customer segmentation, because there is no use for such detailed analysis. Maybe in the next 1–2 years there Xiaomi will use customer segmentation for marketing.

3. **What are the differences in design for Xiaomi watches and bands in the domestic and overseas market?**

There is no difference in design in general. In details, there is a difference in Africa and in India, when it comes to heart rate monitoring. Dark skins have an influence when it comes to use ECG to get the real-time the heart rate. There has improvement on software and hardware design because of skin color differences.

In the beginning, is the design of algorithm based on foreign or domestic users?

This depends on our algorithm and hardware supplier. In the beginning, they did not consider include selling Xiaomi wearables in the overseas market. As Xiaomi is based in China, the designing and testing is focused more on the domestic market and is based on domestic market demand. Xiaomi only expands in the overseas market when the product proves to be successful in the domestic market.

Has Xiaomi explored the business model in the overseas market? For instance, to evaluate the health profile of users with insurance companies and provide users with free or discounted devices?

There has been talks about sharing data with insurance companies not just in the overseas market, but also in the Chinese market. As it involved user privacy issues and sharing data with third parties, Xiaomi is quite conservative about it.

4. **What are the most used features of the Xiaomi band? What are the other user feedbacks on the product?**

The most used function is health related function such as steps and sleep monitoring. There are rarely people who exercise every day. Health covers so many functions, such as steps and sleep monitoring. There are rarely people who exercise on a daily basis. If we put step monitoring under sports related function, then it is the health related function, which is used more often. Initially steps monitoring was considered as part of the sports coaching function, after some internal discussions, it is changed to health function. Walking is no longer taken as a sport.

Step monitoring and sleep monitoring are the two most used functions, with message reading becomes quite popular with users.

There are reasons why payment function is rarely used. Making payments via wearables can happen via QR code or with NFC. NFC only become popular and matures in recent years. Adding NFC payment to wearable devices can incur additional hardware cost. Paying via QR code on wearable devices is difficult because the band and watches have very small screens. Now there is a trend to pay via wearables with some technological barriers. The regulators are generally supportive on adding NFC to wearables.

How to realize swiping Xiaomi bands on the metro station?

It depends on the technology and also policy. Currently, Xiaomi realized payment via wearables with NFC. Now it supports around 200 cities. When it is possible to swipe the public transportation card, it is possible to swipe Xiaomi card. The system has received support from large, middle and small cities, to transfer the functions involved in public transportation card to Xiaomi band. Sometimes at convenience stores where it is possible to swipe public transportation cards, it is possible for users to swipe Mi Band now.

Is it possible for users to monitor health risks continuously? How does Xiaomi deal with the possibility that users will not regularly use wearables?

Users will not just use Xiaomi wearables because of the curiosity. Users usually continuously to use Xiaomi products because the Mi Band is light, with the power can sustain the use of about 20 days to 1 month. Users can continuously to wear it without feeling big disturbance. Therefore the convenience brought by the band outweighs the trouble brought.

When it comes to Xiaomi smart watch, there will be power issue involved. Because sometimes some smart watches need to be charged on a daily basis, or every 3–4 days, or 1–2 days. In the beginning of the design process, there are requirements for power sustainability to go up to more than 20 days. This requirement was implemented to improve usability of wearables in user frequency and habits.

5. **Will it be possible for Xiaomi to add more health related functions to new products such as automatic fall detection, VO2 changes, and to monitor nutrition conditions with cameras?**

   For sure Xiaomi will have more health related functions added to new products in the future. The timeline however is not clear, maybe it will happen after 1–2 year or maybe after 3–5 years.

   There is an entrance barrier for the health industry, with it is difficult for normal wearable producer to go into the industry.

   The strict regulation for the healthcare industry is not convenient for commercialization of health related functions in the wearables. The entry barrier for the healthcare industry is high given the strict regulation, with Xiaomi has the qualification and the scale to go into the industry.

Will the China Food and Drug Administration (CFDA) certification has an impact on the health related functions for Xiaomi wearables?

The health related functions Xiaomi promotes in the future will involve CFDA certification. With heart rate monitoring does not involve CFDA certification. More health related function will involve CFDA certification. Many health related functions for wearables lie in the grey area with many companies are designing their products like this. For instance, the heart rate monitoring function, if wearables just give the user the real-time read of heart rate, then heart rate monitoring remains a health related function which has nothing to do with medical use; therefore, it is not related for CFDA regulation. When it comes to monitor blood pressure, blood sugar, CFDA may get involved. Fall detection function is only related to detect falls, and therefore it is not medical use related. It belongs to the health function.

Is it possible to add the nutrition related data to the health related function, and use cameras in wearables to monitor nutrition for users?

Theoretically speaking, it is good. To realize the function takes time. Although Xiaomi has a hardware based business model, the functions and the products mentioned, if there is enough user demand, Xiaomi will possibly realize it.

What are the other barriers to realize IoT network synergy value?

If there is user demand for every new function, Xiaomi, as an internet company will meet the needs. There is no barriers for realizing the IoT network effects, the question remains whether it is the worth to do it. Some user scenarios are not standard user scenario. Xiaomi is sticking to meeting the needs of 80% of users and their 80% of needs. There is cost related to developing each new function, and this is related to the value of Xiaomi. If only 10% of users need the function, then probably it is not cost effective for Xiaomi to develop such function. It is not that Xiaomi does not realize there is a niche market, it is just because it is not worth to invest in the niche market.

Part 2: Data flow, processing with Xiaomi wearables and associated business model

1. **What third party devices are Xiaomi wearables connected with? Is it possible to link the sleep monitoring data from users and smart home appliances to create better sleep experience for users?**

   Automatic waking up with lights is possible and this solution can share with performers in the market.

   The lights Xiaomi treat them as partners (manufactured by Xiaomi ecosystem firms). Third party appliances are manufactured by other firms other than Xiaomi. For Xiaomi ecosystem appliances, we are trying to improve the IoT user experience, and to win user trust. In case we failed to improve on user experience, Xiaomi will end up losing user trust. In the beginning, the Xiaomi IoT platform is not that open to third party apps. Internally, the IoT network involves lights and air-conditioning. It targets customers who have higher requirements for sleep conditions; the air-conditioning and the fan can automatically switch to sleep mode when detecting users switch off the light. The IoT platform is not open to third party appliances and apps.

How did Xiaomi incorporate Alipay on wearables?

Alipay provides Xiaomi with solution packages, with Xiaomi proposed to incorporate Alipay on wearables. Alipay is relatively open to this types of solution, not just for Xiaomi.

Why did you not cooperate with Wechat?

Wechat is rather conservative in this sense, with all the wearable devices not support wechat payment function. It is also related to how wechat choose their wearables collaborative partners.

The openness of IoT device platforms for smart home appliances differs in the domestic market. How is the openness level for the IoT industry?

All platforms are boasting their openness for other brands of devices, but in reality, none of the platforms are doing a very good job. Xiaomi is not bad in this sense. Except for Xiaomi, all other platforms are in the start-up phase, and are promoting their openness, and try to get more users and manufacturers to use their platforms. In the future, the platform is not mature enough, with their prospects of opening to other brands limited, and this is related to their corporation management.

If users buy different types of smart home IoT devices, can all different types of home appliances get connected to the single IoT platform?

This is related to the historical development. Judging from the current situation different IoT platforms belong to different brands; these brands have no incentive or reason to build inter-connectivity. In the future, if users want better experience on inter-connectivity, Xiaomi will try to stay open to home appliances from other brands. Now it seems difficult.

The IoT system will evolve with the Matthew effect; with good platforms will become more inter-connected with other brands and bad platforms will gradually disappear. The IoT collect with the downstream users and the content and service providers in the upstream. Only bigger platforms with mature technology and services can connect with more users, with more users attract more services and contents. In the end bigger platforms will attract more users and the market will be of competitive oligopoly.

In the end, if users buy smart home appliances from multiple brands, these different appliances may not be able to connect to a single platform. Unless the Chinese government requires for such interconnectivity and all brands answer the call, it is difficult to have such interconnectivity.

2. **Where are the steps, heart rate, and GPS data processed? On the device or in the cloud?**

   Xiaomi use both methods to process data. Basic data is processed on the device; take power sustainability into consideration, complicated data cannot be processed on the device. More complicated calculations are processed in the IoT platform. For users all data are quite sensitive, and Xiaomi tracks data flow to the cloud and on the device.

3. **How does behavior tagging work in Mi Fit? What was data connected for? Does Xiaomi share the data with third parties?**

   The tagging aims to let improve the algorithms and adjust it according to users' needs. In the future, the algorithm can offer personalized solution for users but now it is not that competent. Xiaomi is very conservative about sharing data with third parties, let alone the tagged data.

4. **Does Xiaomi inform users about sharing data with third parties? For instance, the "Steps taken per day" Health insurance, will users share data other than bio-metrics data?**

   Xiaomi will give users notification about privacy policies when users intend to choose certain services which involve sharing data with third parties. If users refuse to accept the privacy policy, then Xiaomi will not provide the services.

5. **Does the user have the right to decide the length of their personal data retaining, sharing parties and the types of data sharing?**

   Because of GDPR and the Chinese privacy protection regulation, users can determine the length of personal data retaining, sharing parties and the types of data shared.

   The Chinese government has implemented the privacy protection clauses. Now users can communicate with Xiaomi via email, calls, for personalized data request such as deleting data on the server. If users fail to communicate or fail to raise personalized request, Xiaomi will deal with the data with standardized solution. Below is the standardized solution offered by Xiaomi (Source: Xiaomi.com, https://www.mi.com/global/about/privacy/, last access June 19th 2019). The standardized solution is what GDPR requires in Europe, and the data privacy law requires in China. The approach Xiaomi takes is the industrial standard approach.

INFORMATION NOT REQUIRING CONSENT

Xiaomi may share anonymized information and statistics in aggregate form with third parties for business purposes, for example with advertisers on our website, we may share them trends about the general use of our services, such as the number of customers in certain demographic groups who purchased Certain products or who carried out certain transactions.

For the avoidance of doubt, Xiaomi may collect, use or disclose your personal information without your consent if it is and only to the extent it is allowed explicitly under local data protection laws.

6. **After the implementation of the General Data Protection Regulations, what are the changes to the privacy policy of wearable devices? Does it affect the business model of wearable devices?**

We cannot answer this question.

Through the analysis of the privacy policy of Xiaomi website , Xiaomi does not sell the user's personal information. Xiaomi will share data with third parties and eco-system companies. The scope of third parties involved in data sharing includes communication service providers, data centers, data storage service providers , advertising and marketing service providers . At the same time, Xiaomi did not disclose data sharing with the eco-system businesses.

According the the Xiaomi Privacy policy webpage, the follow changes has been made.

What is new under GDPR?

"Xiaomi has added Data Protection Office r to ensure GDPR compliance (1) Xiaomi set up a Data Protection Officer (DPO) in charge the data protection, and the contact of DPO is dpo@xiaomi.com; (2) procedure like data protection Impact assessment (DPIA).

Xiaomi updated the types of personal information that we collected and the purposes of collecting such information. For example, we collected hardware usage information to conduct statistical analysis and optimize the performance of your devices.

By complying with GDPR and providing better data privacy protection, we updated the relevant content about users' rights under GDPR, and how we process the personal information for our Europe Union users.

Xiaomi updated the relevant content of third parties' products and services which may be involved during the use of our products and services."

The impact on business model for Xiaomi wearables

GDPR does not have big impacts on the Xiaomi wearable business model, with the core of the business have not changed. The hardware based business model did not change. Data related business belong to derivative services in the later stage. Xiaomi has not started exploring data based business model. There is no existent data derived business from Xiaomi.

7. **What is the added value services targeted Mi Band and Mi Smart Scale?**

All the services now provided are included in the price of the hardware. There is no additional service packages which requires users to pay. The membership based value added service is in the stage of exploration stage with no fixed conclusion yet.

Will Xiaomi analyze data from Mi Fit and Mi Home together and evaluate health risks of users?

This is a possibility of the direction where we are going, and currently there is no such schemes. We cannot answer whether it is in the development stage. The important thing for us to get the data, analyze and to present it to users.

Will Xiaomi improve the interaction methodology with users?

Besides controlling with smartphones or via voice assistance, the other interaction methods such as via eyes, hand movements, or brain waves, will not happen in the near future. Maybe it will happen after 10 or 5 years.

8. **Smart blood pressure monitor (i-Health), smart glucose monitor (i-Health), smart urine analytics reader (i-Health), can all connect to Xiaomi Mi Home app. Will Xiaomi consider take use of the data from these smart health monitors and build a user smart health profiles?**

I cannot answer this question, as this is a problem of our third party partners. Xiaomi has not control of it. Under the smart health umbrella, Xiaomi would like to analyze data in a comprehensive manner. It is not very certain whether Xiaomi can do it. Some data would be very useful for us, but data collection shall happen within the data protection compliance framework. Xiaomi will strictly follow GDPR or the Chinese data protection policy, with the Chinese government also has strict regulation about bio-metrics data protection.

9. **Is it possible to establish patients' personal health records and share them with doctors to provide personalized health coaching? Is this business model applicable in China?**

This proposal sounds very bright. It is difficult in China. In China, the medical resources are very scarce. In the future, Xiaomi might follow such strategy; in the short term, it is very difficult.

10. **The "Steps Sharing Insurance" allows Xiaomi to share steps users taking per day with Zhong An where users can use the steps to trade for insurance premiums. Besides steps taken per day, what other data does Xiaomi share with Zhong An?**

By the end of 2018, Zhong An has collaborated with many tech groups for the "Steps Sharing Insurane" scheme, allowing users to book insurance plans via 51Yund app or Mi Fit App. On data sharing level. Xiaomi has not shared data with Zhong An or other insurance companies. Zhong An can sell insurance via smart fit apps, that is all.

From internet, among the Steps Sharing Insurance users, 47% are born after the 1990s, 37% are born after the 1980s. Young people have low risks to get fatal diseases, and thereby bearing less risks than other groups of the population and enjoy less premium. The premium of such insurance cost less than other fatal disease insurance plans.

11. **Will Xiaomi consider obtain data from smart rice cooker, and blood pressure monitor, smart blood sugar monitor, and other smart wearables, and analyze the data in a comprehensive manner to provide nutritional and sports advice to users?**

In the future it is a plausible scenario. In this stage of development, it can be difficult. Xiaomi will not start such practice now. There are no many technical barriers; the market is not mature enough. Besides, the ECG function embedded in Apple Watch only obtained the FDA certification in the US. The ECG function cannot be used in China.

12. **Does Xiaomi has plans to promote the blood pressure monitoring via PPG to reduce the disturbance to users via blood pumping?**

The method is still in early development stage (by contrast, such products have emerged in CES Asia 2019). The PPG monitoring is not fit for the current market demand.

13. **What is the view of Xiaomi for the fall detection and prevention function embedded in wearables?**

The function is merely a slogan. Users may find it attractive in the beginning, to realize such function is difficult in reality. It involves collaboration with network provider or the issue of data sharing.

14. **Is it plausible for the business model where insurance companies pay for wearable devices?**

Xiaomi has no problem with such business model. However, the standardization to measure the effectiveness of wearable devices is an ongoing process. It is difficult to measure the value of wearable device in reality, therefore not possible for insurance companies to refund users.

CHAPTER 3

Use of AI in Combating COVID-19

Practices in Different Economies

1. Introduction

To combat the Covid-19 pandemic, AI has been widely used to analyze outbreaks, identify close contacts, and develop vaccines and drugs. With the vast amount of capital flowing into the healthcare system, many countries have rushed to develop AI-assisted healthcare monitoring, diagnosis and treatment system. With AI becoming more powerful, tech companies have gathered more and more power with data accumulated within a small number of companies. The use of AI has been extended to the monitoring of covid-tests and travel history — as some refer to as digital handcuffs (McMorrow and Leng 2022). Digital healthcare monitoring equipment has been used for not only pandemic related prevention and monitoring, but also primary healthcare and assisted diagnosis online. Tech companies have marched from online to offline and started to control more and more healthcare resources. To evaluate the impact of such measures, this chapter seeks to discuss the position of different stakeholders in implementing telehealth solutions.

COVID-19 has posed great challenges for the unprepared public healthcare systems with an aging population. The sad truth seems to be that countries with an older population of pre-existing chronic disease conditions (hypertension, cardiovascular disease, diabetes, etc.), such as Italy, Spain, the Netherlands and the United Kingdom, reported higher mortality rates than countries with a younger population, such as China (Johns Hopkins University 2020, Wu et al. 2020). The parts of the population which were hit hardest by the pandemic were the poor and the elderly. Public healthcare systems lacking hospital beds, intensive care units and qualified and trained medical staff can use smart health solutions to resolve the challenges encountered in combating COVID-19 (Philips 2020). Smart health solutions can potentially improve prevention, diagnosis, treatment and recovery efficiency by promoting data accuracy (Hollander and Carr 2020). In a strained healthcare system in post Covid times, getting infected with Covid can lead to increased risk of cardiovascular diseases, changes in brain structure, driving the healthcare system to another strain; this forces medical professionals to re-think about putting preventive healthcare at a higher priority, utilizing telehealth solution to combat the problems brought by the double strikes of Covid and an aging population (Neville 2022). Even though Covid-led digitalization of the healthcare system may break silos between hospital departments and improved data interoperability for many institutions, the resources left for patients with non-Covid conditions, especially patients with chronic diseases, declined significantly.

Three years have passed since Covid-19 first broke out in China and subsequently spread across the globe. Many countries have moved on and started living with the virus. In the recent Covid re-opening in China, big cities like Beijing find themselves trapped in a moral dilemma; the sudden reversal of zero-covid policies has led to a shortage of medical resources such as medications, testing kits and medical staff, along with supply chain disruptions. Online consultation has grown exponentially by 198% as patients are afraid of going to hospitals for the fear of cross infections.

The World Economic Forum has listed the 7 biggest breakthroughs in the healthcare industry, with 'AI can detect skin cancer better than a doctor' and 'Your phone will know if you are depressed or not' ranking 3 and 4 respectively on the list[1] (World Economic Forum 2019). Exciting innovations in smart health industries include utilizing telemedicine to transfer care to a home setting, deploying AI to reduce the workload on physicians and divert patients to the right doctor, and using Internet of Things to improve patients monitoring and coaching; these innovations have left us with the feeling that the healthcare industry is on the verge of an AI revolution.[2] Huge amounts of public and private sector investment have poured into big data and cloud computing, and utilizing such tools in the healthcare continuum. AI seems to promise to bring high value healthcare services and products, possibly ending our quest for the impossible healthcare trinity of access, quality and cost.

---

[1] Robin Pomeroy, World Economic Forum, "These are 7 of the most exciting breakthroughs in healthcare today", published on May 8th 2019, available at https://www.weforum.org/agenda/2019/05/healthcare-technology-precision-medicine-breakthroughs.

[2] Emanuel EJ, Wachter RM. Artificial Intelligence in Health Care: Will the Value Match the Hype? JAMA. Published online May 20, 2019. doi:10.1001/jama.2019.4914.

This book aims to find out whether AI has achieved this aim, with empirical evidence from China and selected European economies. Chapter 3 consists of 5 parts, wherein part 2 deals with concept development, part 3 is about research methodology, part 4 showcases stakeholder power mapping results, and part 5 concerns conclusions and implications.

2. COVID-19 Pandemic and its Economic Impact

The COVID-19 pandemic spread globally at a record speed compared to SARS in 2001. Globalization has deepened our connections, making it easier for people to travel and live on a global scale. This renders it difficult to manage the public healthcare crisis on a traditional national basis. Historical health crisis such as SARS and MERS keep reminding policy makers that global warming will make pandemics such as COVID-19 spread more easily (World Health Organization 2003). However, the world still lacks international cooperation and resolution to deal with a global crisis. Regarding management of the crisis, among major economies, developed economies in Asia have proved to be more successful than developing economies in managing Covid. This was largely because of the closure of borders; for instance, South Korea, Australia, Taiwan and Japan had a lower death rate than China. Because of locks and early vaccination, Asian economies generally performed better than European economies or even North American and South America countries (See Figure 3.1).

The COVID-19 pandemic hit the global economy, with strict quarantine methods enforced all over the globe. Public healthcare systems in certain economies such as Italy and the United Kingdom were not prepared of that. This resulted in high death rates in economies with aging populations (see Figure 3.1). The drama caused by the unavailability of ventilators proved that the current healthcare system remains unprepared and ill-equipped to deal with such a crisis. Except the limited resources—both material and human resources—in the healthcare systems, the high death rates among vulnerable groups remains a highly debated topic. It has been clear that elderly with multiple chronic diseases have been identified as the vulnerable group in the pandemic (see Figure 3.2, Figure 3.3, Figure 3.4, Figure 3.5, and Figure 3.6).

Other healthcare system issues revealed in the pandemic include lack of coordination among different stakeholders, treatment and prevention measures, with the vulnerable groups not properly informed of risks associated with the disease and the related risks. In the end, the vulnerable groups suffered tremendous losses. Most notably in China, the low vaccination rates among the elderly led to a lock down of three years, huge economic costs, as well as relatively high hospitalization rates for those above 60 years of age.

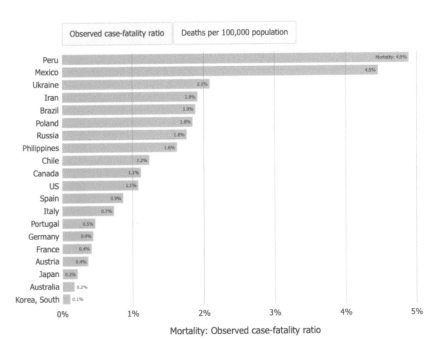

Figure 3.1: Covid-19 mortality rate (death per 100,000 population) by country. Source: John Hopkins University 2022, accessed on March 12th 2020.

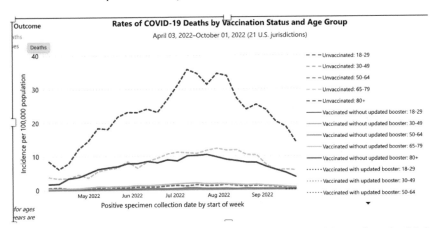

Figure 3.2: Weekly Death counts in different age groups in the United States, from April 3rd 2022 to Sep 25th 2022, CDC NCHS, Last accessed on Dec 2022.

In major developed economies, the general practice of vaccination is that vulnerable groups, such as people with multiple chronic diseases and over 50 years of age, were invited to receive vaccinations first. Such policies make sense for two reasons. First, the chances of hospitalization in the case of Covid for these vulnerable groups were much higher than for the rest of

COVID-19 Weekly Cases per 100,000 Population by Age Group, United States
March 01, 2020 - December 03, 2022*

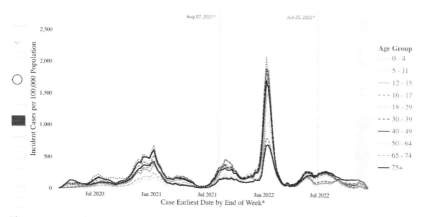

Figure 3.3: Total deaths sorted by age group in the U.S., CDC, Last Accessed on Dec 12th 2022.

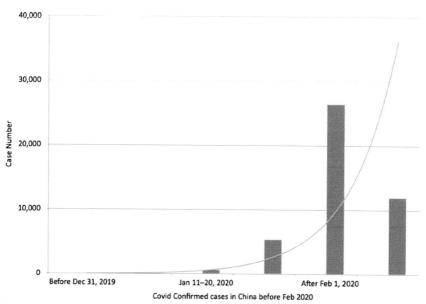

Figure 3.4: Confirmed Covid cases in China before Feb. 2020. Source: Center for Disease Control and Prevention, 2021.

the population. Secondly, Covid vaccines serve to prevent hospitalization rather than total immunization. In China, however, the whole vaccination process remained voluntary—regardless of age, order or chronic disease history. In fact, in many cases, older groups with multiple chronic diseases

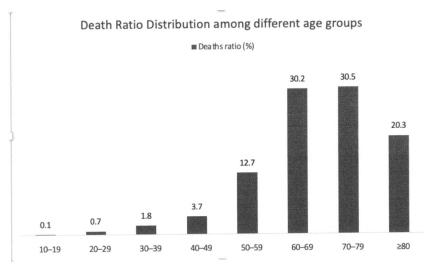

Figure 3.5: Death distribution among different age groups. Source: Center for Disease Control and Prevention, 2021.

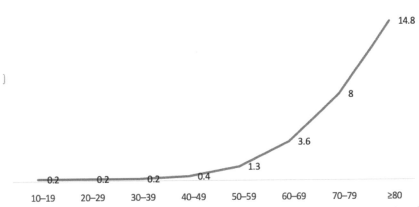

Figure 3.6: Case Fatality Rate by age group. Source: Center for Disease Control and Prevention, 2021.

were either willingly avoiding vaccination, fearing the risks of side effects, or refused vaccination because of their chronic disease history.

Currently, there are four types of globally recognized vaccines — Protein Subunit, RNA, Non-Replicating Viral Vector and Inactivated — developed by 10 manufacturers to combat Covid-19 (WHO 2022). China has developed inactivated vaccines (Sinopharm, Sinovac and CanSino) and drugs for treating COVID-19. In fact, China, as a member of COVAX, remains a net exporter of vaccines, exporting 32.6% of global vaccines, thus ranking second in the world (WTO 2022). Medical device companies such as Fosun have imported fee-based vaccines from abroad and inject the shots in Hong Kong, as the Chinese government allows only China-made vaccines such as Cansino and Sinovac to be used in mainland China.

Only 56.4% of the over-aged (60–69 years age group) in China have received a booster shot. The ratio drops to 48.4% for those aged between 70–79 years (Person and Jones 2022). Moreover, merely 19.7% of those over 80 years of age in China have received booster shots as of March 17, 2022, and just 50.7% of that age group have completed their primary vaccinations.

In most cases, vaccination should be administered by local authorities such as the local China Disease Control Center (CDC) or general practitioners With a weak local healthcare system in place, it is difficult for the Chinese government to effectively identify and vaccinate vulnerable groups. Often a local CDC doctor or family doctor serving at community healthcare centers would cover 6,000–10,000 patients on a yearly basis in their jurisdiction allocation (Chen 2022). Relying on the local level healthcare system alone to provide vaccination services is simply not practical and is difficult as well. However, it is not even possible to divert human resources from hospitals either, where specialists are focused and busy on research, teaching, and treating patients. This explains the chaotic and slow vaccination process for vulnerable groups in China.

Secondly, clinical studies have shown that the deactivated vaccines were less effective than m-RNA-based vaccinations in preventing hospitalization; deactivated vaccines (two doses of Sinovac-CoronaVac vaccine) have been shown to prevent 51% of symphonic infections (Singapore Ministry of Health 2021). WHO suggests that CoronaVac was 51% and Sinopharm 79% effective in preventing symptomatic infections of Covid-19 (Mallapaty 2021). The ratio, compared with the 63% efficacy reported for the University of Oxford–AstraZeneca's viral-vector vaccine, and the 90% and higher efficacies of the mRNA vaccines developed by Pfizer–BioNTech and Moderna, can lower the trustworthiness of Sinovac vaccines in the already vulnerable groups who are not willing to take the vaccines because of trust issues.

Another important point to note is that WHO recommends a third dose for the inactivated vaccinations to work. Otherwise, the effect of such vaccinations wears out rapidly (Mallapaty 2021). Clinical studies of over 1 million people in Brazil who were vaccinated with CoronaVac show that the vaccine offered up to 60% protection against hospitalization for people up to 79 years of age. When it comes to the over 80 years old age group, CoronaVac was only 30% effective at preventing severe diseases and 45% effective against death, compared to 67% and 85% respectively for the Oxford-AstraZeneca jab. With the vaccination ratio of a third (booster) shot low among vulnerable groups, it was difficult for the Chinese government to open its borders as such policies would cause deaths among the elderly at a rate faster than the government could control.

With the Covid-19 variants, many countries, including China and India, continued on with living with the virus (International Monetary Fund 2020); With lockdowns and other travel restrictions disrupting the supply chain and the value chain, the economic projection for the Eurozone was adjusted to be 3.9% for 2021, bouncing back from –7.3% in 2020 (European Central Bank 2021) and for the economic projection of U.S. in 2020 was adjusted to –2.3 % and in 2021 4.2% (United States Bureau of Statistics 2021, The Federal Reserve 2020). Martin Wolf (World Bank 2020) commented on the drastic economic cost of COVID-19 as unprecedented. It is urgent for the world to find new economic growth leverages and balance among the U.S. China trade war and the shocks brought to the global value chain. Whether AI can become the driver for new economic growth and to what extent it will have an impact on the healthcare systems, these questions are widely debated today.

For China, the zero-Covid lockdowns were carried out against huge economic costs. Ever since the lockdown in Shanghai, there had been a manpower shortage in hospitals and care homes in Shanghai as doctors could not leave their homes for work. Shanghai, under normal circumstances, is equipped with the most abundant healthcare resources in China. There are 38 level III hospitals, with top hospitals such as Huashan and Zhongshan hospitals affiliated with Fudan University. Fudan University hospitals often conduct world class research, including sharing the DNA patterns for the coronavirus at the beginning of the pandemic. Despite the abundant healthcare resources, manpower shortages were generally detected in hospitals and care centers in Shanghai because of escalating Covid cases, as well as the mounting numbers of inpatients and outpatients.

During the zero-Covid lockdowns in Shanghai, elderly patients infected with Covid were reported to work as nurses, putting bodies in the morgue at care homes (Xue 2022). For millions of Shanghainese, getting food had become a challenge, ironic for a city with thousands of world-class restaurants. People could not go to the supermarkets as there were not enough delivery workers to serve the 20 million people across the Huangpu River. Shanghai had effectively become a war zone before the lockdown as people were sending their families out and emptying the supermarkets to scramble for food. Trucks that connected Shanghai to the other parts of China, delivering food, were stuck on the road as they could not reach Shanghai, nor could they leave, resulting in food rotting on the way and drivers stuck on the road, living in the trucks as mobile homes.

For the global shipping industry, the lockdown in China added further pressure to the supply chain constraints and stresses. According to

Person and Jones (2022), the delays at the port of Shanghai were worsening. The manpower crunch for ports was caused by workers and administrators who could not leave home for work. Shipments from China to the U.S. increased by 64 days as a result; the number of ships awaiting at the Shanghai port also increased by 34% on a year-on-year (YoY) basis to 344. In Europe, shipments from China were reported to be late by 4 days on an average. The aggregate time of turnaround increased by 8% in Rotterdam, 30% in Antwerp and 21% in Hamburg. The global fuel cost also increased on a YoY basis of 66% in Singapore.

For the Chinese government, which put the country under quarantine and carried out the strictest travel policy for 2 years, the current situation marks a clear public health policy failure. The ultimate cause and result of the zero-Covid policy is that it created a false sense of safety, making the country and its 10 billion people forget that Covid remains a global healthcare challenge. The resulting supply chain disruptions will last for years as the world diversifies supply chain risks fron China, and businesses are considering relocating to other locations where policies are more predictable; moreover, the economic development model powered by heavy infrastructure development and local government land sales is not sustainable given the three-years of disruptions of Covid (Feng and Li 2023, Chen 2016). For the rest of the world, it is likely that inflation will continue. Companies will adjust the price tags and make consumers pay for higher shipping bills, unreliable supply chain and longer waiting times.

For Chinese citizens, the government's control over the citizens, in the name of Covid, has increased ten-fold since March 2020. The nationwide use of a QR code-based monitoring system has prompted each province in China to develop its own version of a Health kit; the health kit allows the monitoring of the daily travel routines of its citizens and their close contacts. Although similar schemes have been developed in other countries such as the Netherlands and the United States, the system was only short-lived and eventually stopped because of the high vaccination rates in those countries. In China, however, the system was sustained at a scale of massive surveillance by local governments to city streets and village level authorities.

The former PBoC head, Zhou Xiaochuan, warned that China is heading back to an old route of closure, and deglobalization trends and nationalism thoughts are becoming popular with many Chinese citizens (Tang 2020). The former PBoC governor worries that this will hurt the Chinese economic growth and threaten the global trade system which China benefits from and relies on extensively. Although the zero-Covid policy ended abruptly after street protests, billions of people were infected with the virus suddenly, causing medical resources shortages in terms of both manpower and medication availability. The sudden lifting of all measures of control in terms of travelling inside and outside China, despite low vaccination rates among those over 70 years of age as well as the low vaccination effectiveness rate, suggests that the Chinese government has grave concerns regarding the economic consequences of the zero-Covid policy (See Figure 3.7 for IMF projections).

More importantly, as illustrated by Table 3.1, the number of public hospitals in China has declined from 2018 to 2022—from 12,032 to 11,804, by 1.89% (National Healthcare Commission 2019, 2022); The number of rural clinics and hospitals at the township level has dropped from 36,461 to 34,463 (by 5.48%); whereas the number of village level clinics has declined from 622, 001 to 599,292 (by 3.65%). Furthermore, the number of specialized care facilities for prevention healthcare, women and children care has also dropped from 18,033 to 13,276 (by 26.37%). While the number of private hospitals has grown from 20,977 to 24,776, the trust in private hospitals has not improved. Access to healthcare services in China has deteriorated rapidly at every level (city, township, village) with Covid squeezing the local government's financial resources, thus making

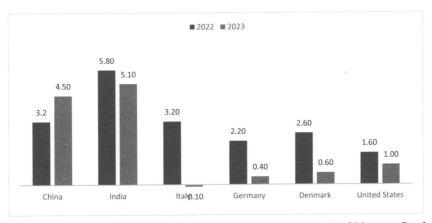

Figure 3.7: World Economic Projection Visualization. Source: International Monetary Fund last accessed on Dec 27th 2022.

Table 3.1: Medical resource distribution in China. Source: China Health Commission 2021.

	Public Hospitals	Township Clinics	Village Level Clinics	Specialized Care Facilities (Prevention, Women & Children care, Birth Control, Supervision)
2018	12,032	36,461	622,001	18,033
2021	11,804	34,463	599,292	13,276
Percentage of Change	−1.89%	−5.48%	−3.65%	−26.37%

it difficult for the public finance system to fund public health initiatives. Although online hospitals and pharmacies filled in, supply chain issues were mounting as truck drivers and delivery workers also caught Covid.

Moreover, a thorough examination of the ownership structure of privately owned hospitals in China, such as Amcare Women's & Children's Hospital, Beijing, reveals that tech companies like Bytedance (company behind Tik Tok) have started to extend their footsteps into the digital healthcare industry. Other companies such as Tencent have also started exploring the Internet + Healthcare model, with the aim of making healthcare more efficient and less costly. However, trust in internet-based hospitals has not been established since the business model often focuses on online-offline medication sale. Patients in China still trust public hospitals more than they trust online diagnosis and treatment measures. Wechat, however, has become a popular source of medical information provided by the official accounts of public hospitals, private clinics, governments, as well as forums.

As healthcare systems in major developed economies suffer a tremendous shock due to the lack of resources, coordination and material support, because of COVID-19, there is clearly a need to rebuild the resilience of healthcare systems and restore public trust in healthcare service providers. Till today, the cost of healthcare is high in major developed economies, while the salary for healthcare workers remains low. The funding cuts and austere policies make many doctors quit their jobs in healthcare for better life and reward packages in sectors such as consultancy, banking and insurance. The impossible trinity in healthcare — low healthcare costs, improved healthcare quality and improved doctor/patient experience cannot be achieved just with the application of AI; AI will certainly drive the need for additional talents in programming and demand for more mature solutions from tech firms nonetheless. The next part of this thesis will discuss how AI will likely interact with healthcare systems.

3. The use of AI and IoT in Healthcare

COVID-19 has pushed the digitalization of healthcare system. Investments in health-related projects have thus grown rapidly. AI has been deployed for detecting disease concentration and spreading, real-time monitoring, and predicting pandemic outbreaks and mortality risks (Arora et al. 2020). AI has also been instrumental in the diagnosis of COVID-19 by performing image recognition on X-Ray and MRI results. For hospital management, AI has become useful in facilitating resource allocation by automating resource management and supply chain management, assisting staff training with VR and AR, maintaining healthcare records, and identifying patterns for trend recognition (Arora et al. 2020).

In pandemic tracking and prediction, AI has been useful in collecting data from social media and identifying disease clusters. In the beginning of the pandemic, Bluedot reported the disease cluster of pneumonia cases in Wuhan by analyzing news reports on Dec 31st, well ahead of the public health administration agencies in China and other economies (Khan 2020). The Johns Hopkins University Coronavirus Resource Centre collects publicly available information and visualizes the data, making it possible to actively track the spread of the disease (Johns Hopkins University 2020). It is now also possible to use Google Maps to estimate active COVID-19 cases country-wise (Banerjee 2020).

With the zero-Covid policy in China, making in-person visits to hospitals has become increasingly difficult for patients. For doctors, the number of patients in both inpatient and outpatient departments has grown exponentially. A shortage of medical resources has been observed at all levels, given that the number of public hospitals and specialized care hospitals have declined during Covid. The public finance system is now overwhelmed by the increasing costs of maintaining the zero-Covid policy, such as testing.

For contact tracing, U.S. universities such as MIT and Harvard have been developing and using contact-tracing apps such as Safe Paths (Safepaths 2020). Tech companies like Google and Apple are also working together to develop contact tracing APIs (Apple 2020). During the pandemic, mobile apps were developed quickly by different governments and tech companies to facilitate contact tracing, with Wechat in China and Coronamelder in the Netherlands for instance. Apps such as AI4COVID-19 have enabled detection of COVID-19 with 3 seconds of coughing and deliver the diagnosis within 2 minutes (Imran et al. 2020).

For the early diagnosis of COVID-19, algorithms were rapidly developed and deployed by tech companies in China to identify patients with symptoms of COVID-19. After the Chinese New Year, when large-scale infections were reported, Yitu Technology developed the algorithms to facilitate the diagnosis and treatment of COVID-19 (Johnson and Johnson 2020). The software received clinical approval quickly and was deployed first in Hubei and then nationally in healthcare systems where the pandemic hit the hardest. The AI assistant received approval from the healthcare service staff. Furthermore, congestion in hospitals was relieved with the deployment of such systems, with patients diverted to infectious disease hospitals. Yitu aimed to establish the AI-assisted paradigm in four stages of the pandemic control process. In the prevention stage, chatbots and online consultation could educate users and have users perform self-examination. In the quarantine stage, the system could help doctors monitor patients and manage their conditions. In the patient screening stage, the system could guide patients to the right medical path. In the diagnosis and treatment stage, the algorithm could facilitate the diagnosis of COVID-19 from 2 minutes to 20 seconds, with the classification of CT images, thereby lowering the cross-infection risk for patients during quarantine/treatment at home. The AI assistant also marked the areas in the lungs impacted by COVID-19, thereby making it easier to track a patient's progress. Researchers around the world have delivered at least 5 other neural networks to diagnose COVID-19 patients (Singh et al. 2020, Alom et al. 2020, Li et al. 2020, Soares 2020, Farooq et al. 2020).

In patient management, AI has been deployed at hospitals to automate asset management (Huawei 2021) and prioritize COVID-19 patients in ICU units for ventilators. AI can also predict the possibility of patient recovery and mortality by monitoring the patient's daily electronic health records and helping doctors to make decisions regarding the next step (Arora et al. 2020).

In pharmaceutical development, AI can accelerate drug and vaccination discovery by reducing the time required for drug discovery, virtual screening and validation processes (Arora et al. 2020). Researchers have quickly obtained genetic information from patients and offered it to the international community (University of Sydney 2020). AI has made it easier to predict the structure of protein (Callaway 2020). This has given the pharmaceutical companies an opportunity to develop the vaccination for COVID-19 quickly. For instance, AI has been used to develop m-RNA vaccines by Oxford University and Moderna (Harvard Business Review 2020).

The concept of connecting the healthcare system with the smart city system has been raised ever since the birth of the concept of Internet of Healthcare Things. COVID-19, being an air-borne disease, had prompted a new normal globally—online education, virtual meetings and remote monitoring and prevention in healthcare systems. The need for connecting smart buildings and smart transportation systems with healthcare systems now seems more urgent than ever.

As suggested by the study conducted by Siemens (2020), the smart building system can play a significant role today in attracting and retaining talents. The reason lies in the fact that employee wellness is now being given more emphasis than before. Retaining employees, given the high mobility of millennials, has become a priority for many multi-nationals. Features such as automatic lighting system and window shades can create a safe and comfortable working environment for employees. Meanwhile, the smart quarantine and social distancing rules implemented by many governments around the globe considerably removed the difficulty in maintaining a safe working space. A smart building management system can contribute to maintaining social distancing, thus protecting the safety of employees (Bloomberg 2020, Siemens 2020). The smart building system can also help monitor the air quality for tenants, thereby prompting the tenants to seek better air quality from landlords.

This book seeks to answer the question whether smart health solutions can perform as they promised, by analysing answers from stakeholders in China and selected countries in Europe.

a. The digital healthcare readiness in European economies

To evaluate the implementation of telehealth solutions in Europe, there is a need to look for the enablers of digital health solutions, with case studies in specific European economies. The case study is conducted on the digital readiness for combating diabetes in Europe, with the 10 European economies. Considering the vast differences within the European healthcare systems, it is necessary to address the issue with a technical framework for legislation, policy, disease guidelines, and patient refund for telehealth solutions.

The Economist Intelligence Unit (2020) released a study regarding the readiness of the European economies for digital healthcare. The study analysed the use of digital health solutions to manage diabetes in 10 European economies. There were five enablers identified for the implementation of telehealth solutions for the management of diabetes, the most important being the availability of reimbursement pathways, followed by the openness of the evaluation and assessment processes for digital healthcare management tools. It is also important to incorporate digital health management tools in disease treatment guidance plans for diabetes. Meanwhile, there is a need to provide training to healthcare professionals to raise their awareness of digital health tools. Finally, turning digital healthcare policies into reality requires time, funding and political commitment.

Type I and II diabetes are a growing global healthcare issue. The reasons for evaluating the digital capabilities for diabetic treatment is the significant number of patients with Type I and Type II diabetes, the huge economic cost, and the impact of real-time monitoring of glucose level. The number of such patients has grown from 151 million in 2000 to 463 million in 2019 (Economist Intelligence Unit 2020). In Europe, the prevalence of diabetes is projected to reach 65 million by 2030. Meanwhile, the healthcare expenditure on diabetes was 229 million in 2019. With digital healthcare tools, it becomes more convenient for patients to track their health status, manage the disease together with doctors and lower the cost of treatment simultaneously.

The aforementioned study was conducted from several perspectives, such as readiness for digital healthcare for diabetes, and fitness of incentive and payment systems. The study incorporated northern, southern, western and eastern European countries, including Belgium, Denmark, England, France, Germany, Italy, the Netherlands, Portugal and Slovenia.

With real-time monitoring of glucose level and data sharing, patients could improve sleep quality and well-being, thereby reducing severe hypoglycemia (Polonsky and Fortmann 2020).

A score was developed based on the following questions regarding Covid-19 response: "Is there a national eGovernment policy?", "Is there a national eHealth policy or strategy?", "Is there a national health information system (HIS) policy or strategy?", "What are the funding sources for eHealth?", "Are key legal eHealth frameworks like jurisdiction, liability and data privacy present? (None, Some, Most, All frameworks)", "Is there a national EHR system?", "Is there legislation governing the use of the national EHR system?", "Which types of health facilities are using EHRs? (Primary, Secondary, Tertiary)"," Is there national infrastructure for remote patient monitoring?", "Is there a national strategy to ensure system interoperability within the health system?"," Is there legislation governing the sharing of data among private companies (e.g., device manufacturers) and government?", "Is there a policy or strategy governing the use of big data by private companies?", "Is there eHealth training for health professional pre- and post-qualification?", "Are digital tools included in health technology assessment (HTA) for medical devices?", "Are digital tools included in reimbursement pathways for medical devices?". Based on these questions, the surveyed European economies were ranked and compared. It was observed that Germany, Italy and Denmark ranked high in the overall score for digital readiness regarding the management of diabetes during COVID.

Among the surveyed economies, another part of the study focused on the readiness of the European economies for digital diabetic care. The following questions were asked: "Is there an operational policy/strategy/ action plan for diabetes?", "Does the national diabetes plan include digital diabetes?", "Is there a national diabetes registry?", "Are digital diabetes tools recommended in national clinical guidelines for diabetes?", "Are relevant healthcare professionals trained on digital diabetes tools? (None, Some, All relevant HCPs)". For the "Are digital diabetes tools recommended in national clinical guidelines for diabetes?" question, only Spain, England, Italy and Germany responded yes.

In Demark, many patients were initially worried about the management of diabetes if infected with COVID. Due to the strong digital healthcare infrastructure, Demark was able to introduce an online consultation software for diabetic patients at primary care and specialist care levels. However, since digital consultation was not included in the national diabetic treatment guidelines, the digital treatment was picked up differently across the country (Economist Intelligence Unit 2020).

Germany is a leader in offering reimbursement for telehealth apps (Bamer 2022); yet it lacks a national level EHR system. There are areas where the country can improve its digital diabetic care, such as offering online prescriptions and pharmaceutical services. Most patients still need to go to pharmacies physically to pick up medication and insulin (Economist Intelligence Unit 2020).

The Italian healthcare system appears to be highly regionalized. The connectivity of the healthcare system remains a problem there. Even though digital pathways for treating diabetes have been included in the national diabetic strategy and clinical pathways, access to digital tools remains limited due to regional differences. Reimbursement for most key digital tools is open in Italy. However, during the pandemic, there was no significant difference in the health outcome of the real-time monitoring data for glucose (Economist Intelligence Unit 2020), and the impact of COVID-19 is yet to be observed.

b. Stakeholder analysis for the implementation of telehealth solutions in developing economies: South Africa

The implementation for telehealth solutions in developing economies may encounter different problems than in high-income economies. When analysing the available literature review, it is therefore necessary to address the challenges posed by the healthcare system in emerging or developing economies. In this paragraph, we consider, as an example, the case of South Africa.

Lack of infrastructure and trained medical staff (Eze et al. 2018, Avgerou 2008, Xiao et al. 2013) has been known to be a hurdle in providing healthcare services in developing economies. Telehealth solutions can become a means for providing health interventions in such countries (Chang et al. 2013, Dammert et al. 2014, Mars 2013, Varshney 2014), preventing communicable diseases and improving the health literacy in healthcare workers and patients.

In studying the stakeholder perspectives for the implementation of telehealth (mHealth) solutions in developing economies, the author adopted a meta study methodology and surveyed 108 papers in the area in order to analyse the interactions between different stakeholders for implementing telehealth solutions in developing economies. 65% of the sample papers were from African countries; most of the initiatives in Africa are funded by public-private partnerships, NGOs or overseas initiatives (Istepanian and Woodward 2016). Around 26% of the sample papers came from Asia, while 9% were from America. The stakeholders were put into five categories: patients, healthcare workers, facilitators, knowledge base and system developers. The perspective of each stakeholder was then investigated for its interaction with other stakeholders and among themselves. Take the patient group for example; the interactions between patients and healthcare workers, patients and facilitators, patients and system developers, patients and knowledge base, and patients and other patient groups were investigated in this case.

The meta-study suggested that there is extensive literature on the perspective of healthcare workers; however, there is gap in the study of interactions between patients and other patient groups and, most importantly, limitation in research on the interactions between system developers and users. In the rural areas of developing economies, when patients are trained to care for other patients, peer-exchange can provide support for counselling and information (Chang et al. 2013). Moreover, the interaction with system developers is key to the early detection of problems, and identifying demands and requirements and presenting solutions for complex problems (Brown and Wyatt 2010, Buchanan 1992). The gap identified in literature suggests that there is a lack of design context for system developers to identify the needs and demands of healthcare workers and patients. Therefore, there is a need to create an open-source eHealth platform to collect data in limited resource settings; the collaborative development approach will create an open-source interface allowing and encouraging telehealth solutions to adopt common standards and solve the data interoperability problem, thus making it more cost effective. Leonard et al. (2020) uses South Africa as a case study on analysing the barriers and facilitators for implementing telehealth solutions in resource restricted setting. The study was based on the implementation of a hearing screening device in South Africa for over two years. The study analysed the implementation process at four levels: the community level (individuals), the health provider level (healthcare professionals), the district health system level and the macro health level for oversight.

The study concludes that need-based innovation improves the possibility of implementation. Meanwhile, changing the device language to a local language would improve device interoperability. Still, having a feedback channel, improving communication with community healthcare workers and having protocols to resolve conflicts have been identified as the key challenges for such implementation. Lacking long-term national level political support for the program, along with a lack of dissemination channel is also one of the identified barriers. On the patient level, there is a need to improve health education on hearing and promoting patient mobility. Changing the patients' perceived view on the public health system is also important (Leonard et al. 2020).

For South Africa, the vaccination rate (population receiving at least one dose) remains at 40.4%. Evidence suggests that an average of 65% vaccination rate among OECD countries was able to lower Covid-19 infection rate by 86% (OECD 2022). This suggests that in the African continent, where 99% of Covid vaccines were imported or donated, in combination with a decentralized distribution model, it is difficult for the South African government to distribute vaccines (Nature Publishing Group 2022, WHO 2022).

Compared with the Chinese healthcare system, where most of the high-quality hospitals are public and the heterogeneity of the patient population in terms of culture and language in South Africa means more difficulty to promote data interoperability. The lessons learnt from the South Africa healthcare system are to promote health literacy for patients and healthcare workers, to offer health education for the use of telehealth solutions and to build the education and feedback channels for community health workers.

c. Case Study on the use of AI for Covid-19 related pandemic monitoring in China

To maintain the Zero-Covid policy in China, the Chinese government mainly resorted to mass testing and travel restrictions. There were three types of travel restrictions and lockdowns in China: for travelers from abroad, for local Covid cases, and for travel within China. Health status monitoring was done with a combination of QR code system (red, yellow, green) and travel history monitoring system (red, yellow, green).

The first type of travel ban applied to travelers from abroad. Currently, there are limited flights to and from China. Take the Netherlands for example, there are three companies operating flights from the Netherlands to China: China Eastern, China Southern and Xiamen Airlines (Ministry of Foreign Affairs 2021). The travel restrictions were as such that there was only one flight per week for each airline. Passengers who traveled to China were required to take two Covid tests at a single institution or two separate tests at two different institutions. With the Omicron infection wave in China, some airlines added further requests such as a Covid test within 24 hours before boarding (Xiamen Airlines 2022). The strict guidelines mounted pressure on airlines which were hit because of Covid, given that the already scrambled flights would be canceled for two weeks once positive Covid cases were detected among passengers when they landed. Quarantine measures from 3–4 weeks often applied to passengers once they landed. During the quarantine, each passenger received a red code from the official QR code system, effectively preventing them from traveling inside China. The assigned medical staff conducted Covid tests for each passenger every day once they were put in a quarantine hotel.

The second type of quarantine was applicable to local Covid cases. Once there were local cases detected, the residential compound where the local cases were discovered often went into quarantine. The patient was transferred to a quarantine hospital or facility until he or she tested negative again for 7 days. This rule applied to all patients regardless of their age, the seriousness of their condition or their hospitalization needs. This resulted in many cases with no symptoms or minor symptoms ending up in quarantine; socially, patients were subject to discrimination by their neighbors who were in fear of a similar fate of quarantine and infection, or by the general population who treated the disease as a stigma.

The third type of quarantine was for travel inside China. Depending on the number of cases detected, Chinese cities were classified as low risk, middle risk or high risk areas. The QR code system monitored the travel history of the population. Once the passenger had entered a middle risk or high-risk area for more than 4 hours, the QR code would change to yellow or red, making it difficult for the subjects to travel without further testing or quarantine.

Overall healthcare policy-making in China remains centralized, whereas the implementation of such policies lies in the hands of local governments. This has created an interesting phenomenon whereby the healthcare policies made by the central government are distorted by the local governments; the central government blames the local governments when things go downhill and credit themselves when policies work out.

4. Stakeholder Power Analysis

To combat COVID-19, the stakeholders can be divided into five groups: governments (central and local governments), hospitals (public, private, online, doctors and nurses), tech companies (telecommunication service provider, telehealth solution provider), pharmaceutical companies (public and private), and patients. The stakeholder interests often differ, resulting in non-coordinated efforts between different stakeholders.

Figure 3.8 illustrates how different stakeholders interacted with one another during the COVID-19 pandemic in China. To maintain the zero-covid policy, mass testing in China was done with hospitals in China, cooperating with local governments. While the central government conducted research and decided on the healthcare policy, the local governments oversaw the implementation. The local governments collected information from residents—travel history from telecommunication service provider, medication prescription history from hospitals, pharmacies and online pharmacies (especially cold related medication), and data from telehealth solution providers; they also conducted mass testing, reported infections, decided on high risk, middle level risk and low risk areas, built detention centers for COVID patients, etc.

In terms of power dynamics, the local governments have the right to decide on the type of data to be collected from tech companies, hospitals, pharmacies and individual citizens. The central government oversees the analysis of data and makes instructive policy decisions. When it comes to influence on policy making, governments at both central and local levels have the highest level of influence. Hospitals and tech companies

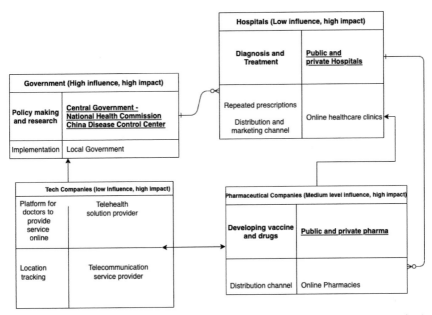

Figure 3.8: Stakeholder mapping for maintaining the zero-Covid policy. Source: Author's illustration.

in China have low level of influence in COVID-related policies, while pharmaceutical companies in China have medium level of influence, given the data generated for vaccination effectiveness. Individual patients have relatively low level of influence in COVID-related policies in China.

As AI was used on a large scale for maintaining zero-Covid policies in China, the level of impact of the policies made by the Chinese government to track the travel history and COVID testing of the citizens was high. The level of impact of tech companies was also significant in China, given that they provided the data to the government—on medication purchase history, travel history, COVID test history, etc. The impact of hospitals was obviously high in terms of patient care and close contact tracking. Furthermore, the level of impact of online clinics and pharmaceuticals has grown during the pandemic as patients prefer online diagnosis and treatment instead of going to hospitals for minor conditions. For pharmaceutical companies, the level of impact has grown during the pandemic with the development of vaccines. In developing economies, the impact is even higher compared to developed economies because of the higher demand for medications and often a lack of sufficient medical resources because of import control. The Chinese government also relied on personal information collection from pharmacies to determine as to who bought cold/pneumonia-related medication in order to track patients.

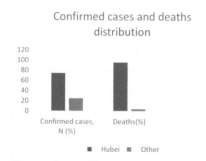

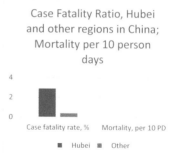

Figure 3.9: Confirmed Covid-19 cases, death distribution, case fatality rate, and mortality per 10 person, December 2019–February 2020. Source: CDC Weekly, 2020.

To showcase how regional policies may differ, Hubei and other regions show a vast difference in COVID-19 case distribution, death distribution, case fatality rate, and mortality per 10 person ratios in the beginning of the pandemic. For instance, the confirmed cases in Hubei region for COVID-19 was about three times as high as other regions (see Figure 3.9), whereas the case fatality rate was over 7 times as high in Hubei compared to other regions. This was because in the beginning of COVID, Hubei went under lockdown, deterring inter-regional travel; except Wuhan, other rural and less developed areas have limited medical resources, resulting in higher death ratios among patients. This resulted in a much stricter zero-Covid policy in Hubei. For instance, travel inside Hubei province was limited and restricted, meaning residents from one city in Hubei needed a COVID test to travel to another city in Hubei.

5. Summary and Policy Implications

Weak local level healthcare system in China and in other developing and developed economies is caused by a lack of family doctors and trustworthy community-level healthcare services. Local doctors often feel a lack of respect and career development opportunities compared to colleagues serving at specialist hospitals. It is difficult to recruit local doctors as well as to design a training system subject to the needs and demands of the local healthcare system. No doctor with 8 years of medical school training is willing to serve at the local level. College students do not wish to choose to be local doctors either. Thus, further collaboration and communication between local governments and the central government over creating an effective healthcare system is needed for a constructive transformation to happen.

The current top-to-bottom policy-making framework in most healthcare systems needs to be adapted to a more bottom-to-top approach. The former slows down the decision-making process, thus causing local-level healthcare crisis and problems to escalate to national and international levels. Effective decision-making framework should be based on local needs and demands, contrary to following orders from central governments.

Local doctors should be paid better and given other incentives such as better career development opportunities to stay. In China for instance, the doctor-to-patient ratio can be lowered to about 1/2000 or 1/3000 for effective local healthcare needs to be met. This would also help resolve the tensions between doctors and patients at level 3 hospitals, while tackling the general dissatisfaction with the healthcare services received as well.

Education regarding the urgency of vaccination is also required for the public. It is particularly important to vaccinate vulnerable groups and improve the vaccination ratio as they suffer the most in case of an infection. China is likely to get a 2G system in place and offer other policy incentives for the public to be vaccinated to avoid further lockdowns and in-humanitarian methods to contain COVID.

Covid lockdowns in China have caused stress in the global supply chain and low economic growth concerns in China and the rest of the world. With de-globalization concerns and the 'slowbalization' reality, China needs to think carefully about its public healthcare policy process by adopting a more preventive healthcare policy mechanism. Although the policy was made because of the poor local healthcare system and low vaccination rates among vulnerable groups, such consequences are causing pain to many communities in China and have trickled and threatened the fragile global supply chain. To avoid future public healthcare crises, the Chinese government should focus on developing and transforming their currently weak local healthcare system.

For the use of AI in monitoring, diagnosis, treatment and digital healthcare solutions, COVID did accelerate the process by driving the demand for telehealth services, and squeezed the healthcare system for manpower. However, access to healthcare services did not improve and the cost of receiving healthcare services deteriorated, suggesting that AI did not mend the gap created by the lack of community healthcare in developing countries and nor could it help attaining the now seemingly unattainable trinity of healthcare.

References

Alom, M. Z., Rahman, M. M., Nasrin, M. S., Taha, T. M., and Asari, V. K. (2020). COVID_MTNet: COVID-19 detection with multi-task deep learning approaches. arXiv preprint arXiv:2004.03747.

Alzubaidi, M. A., Otoom, M., Otoum, N., Etoom, Y., and Banihani, R. (2021). A novel computational method for assigning weights of importance to symptoms of COVID-19 patients. Artificial Intelligence in Medicine, 112: 102018. https://doi.org/10.1016/j.artmed.2021.102018.

Apple. (2020). Privacy-Preserving Contact Tracing - Apple and Google. Retrieved December 16, 2020, from https://covid19.apple.com/contacttracing.

Arora, N., Banerjee, A. K., and Narasu, M. L. (2020). The role of artificial intelligence in tackling COVID-19. Future Virology, 15(11): 717–724. https://doi.org/10.2217/fvl-2020-0130.

Avgerou, C. (2008). Information systems in developing countries: A critical research review. Journal of Information Technology, 23(3): 133–146. https://doi.org/10.1057/palgrave.jit.2000136.

Bally, E. L. S., and Cesuroglu, T. (2020). Toward Integration of mHealth in Primary Care in the Netherlands: A Qualitative Analysis of Stakeholder Perspectives. Frontiers in Public Health, 7: 1–17. https://doi.org/10.3389/fpubh.2019.00407.

Banerjee, S. (2020). Navigate safely with new COVID data in Google Maps, Google. Available at: https://blog.google/products/maps/navigate-safely-new-covid-data-google-maps/ (Accessed: 22 May 2023).

Bloomberg. (2020, July 8). Bloomberg, The Future of Smart Cities: Smart Cities 2.0. Bloomberg's Smart Cities 2.0 Virtual Event. https://onlinexperiences.com/scripts/Server.nxp?LASCmd=L:0&AI=1&ShowKey=97803&LoginType=0&InitialDisplay=1&ClientBrowser=0&DisplayItem=NULL&LangLocaleID=0&SSO=1&RFR=https://onlinexperiences.com/Launch/Event.htm?ShowKey=97803.

Brown, T., and Wyatt, J. (2010). Design Thinking for Social Innovation. Development Outreach, 12(1): 29–43. https://doi.org/10.1596/1020-797x_12_1_29.

Buchanan, R. (1992). Wicked Problems in Design Thinking. Design Issues, 8(2): 5. https://doi.org/10.2307/1511637.

Callaway, E. (2020, November 30). "It will change everything": Deepmind's ai makes gigantic leap in solving protein structures. Nature News. https://www.nature.com/articles/d41586-020-03348-4.

Center for Disease Control and Prevention. (2021, February 11). Provisional COVID-19 Death Counts by Sex, Age, and Week. Retrieved February 11, 2021, from https://data.cdc.gov/d/vsak-wrfu/visualization.

CDC Weekly, C. (2020). The epidemiological characteristics of an outbreak of 2019 novel coronavirus diseases (COVID-19)—China, 2020, China CDC Weekly, 2(8): 113–122. Available at: https://doi.org/10.46234/ccdcw2020.032.

Chang, L. W., Njie-Carr, V., Kalenge, S., Kelly, J. F., Bollinger, R. C., and Alamo-Talisuna, S. (2013). Perceptions and acceptability of mHealth interventions for improving patient care at a community-based HIV/AIDS clinic in Uganda: A mixed methods study. AIDS Care, 25(7): 874–880. https://doi.org/10.1080/09540121.2013.774315.

Cesuroglu, T. (2016, January). Integration of a personalized health care model into health systems and policies in Europe. Maastricht University. https://cris.maastrichtuniversity.nl/ws/portalfiles/portal/754541/guid-d34cba94-a25e-442a-b936-e01facdfadd0-ASSET1.0.pdf.

Chen. N. (2022). Stakeholder power analysis of the facilitators and barriers for telehealth solution implementation in China: A qualitative study of individual users in Beijing and Interviews With Institutional Stakeholders. JMIR Form Res., 6(1): e19448, doi: 10.2196/19448, PMID: 35044321, PMCID: 8811689.

Dammert, A. C., Galdo, J. C., and Galdo, V. (2014). Preventing dengue through mobile phones: Evidence from a field experiment in Peru. Journal of Health Economics, 35: 147–161. https://doi.org/10.1016/j.jhealeco.2014.02.002.

Dijkman, R. M., Sprenkels, B., Peeters, T., and Janssen, A. (2015). Business models for the Internet of Things. International Journal of Information Management, 35(6): 672–678. https://doi.org/10.1016/j.ijinfomgt.2015.07.008.

Economist Intelligence Unit. (2020). Digital Diabetes Index: Enhancing diabetes care through digital tools and services. The Economist Intelligence Unit Limited. Retrieved from https://digitaldiabetesindex.Economist Intelligence Unit.com/index.

Emanuel, E. J., and Wachter, R. M. (2019). Artificial Intelligence in Health Care. JAMA, 321(23): 2281. https://doi.org/10.1001/jama.2019.4914.

European Central Bank. (2021, January 7). Economic Bulletin. Retrieved February 11, 2021, from https://www.ecb.europa.eu/pub/economic-bulletin/html/eb202008.en.html.

Eze, E., Gleasure, R., and Heavin, C. (2016). How can mHealth applications that are developed in one area of the developing world be adapted for use in others? Journal of Decision Systems, 25(sup1): 536–541. https://doi.org/10.1080/12460125.2016.1187414.

Eze, E., Gleasure, R., and Heavin, C. (2018). Mobile health solutions in developing countries: a stakeholder perspective. Health Systems, 9(3): 179–201. https://doi.org/10.1080/20476 965.2018.1457134.

Farooq, Bazaz. (2020). A novel adaptive deep learning model of Covid-19 with focus on mortality reduction strategies, Chaos, Solitons & Fractals, Volume 138, 2020, 110148, ISSN 0960-0779, https://doi.org/10.1016/j.chaos.2020.110148.

Feng, R. and Li, C. (2023). One of China's poorest provinces faces imminent debt problem, The Wall Street Journal. Available at: https://www.wsj.com/articles/one-of-chinas-poorest-provinces-faces-imminent-debt-problem-5ca1becf (Accessed: 22 May 2023).

Gross Domestic Product, 4th Quarter and Year 2020 (Advance Estimate) | U.S. Bureau of Economic Analysis (BEA). (2021, January 28). Retrieved February 11, 2021, from https://www.bea.gov/news/2021/gross-domestic-product-4th-quarter-and-year-2020-advance-estimate.

Huawei. (2021). Overview of the Healthcare IoT Solution - S5700 and S6720 V200R012C00 Configuration Guide - WLAN-AC - Huawei. Retrieved January 25, 2021, from https://support.huawei.com/enterprise/en/doc/EDOC1100038361/c8152cf2/overview-of-the-healthcare-iot-solution.

Hollander, J. E., and Carr, B. G. (2020). Virtually Perfect? Telemedicine for Covid-19. New England Journal of Medicine, 382(18): 1679–1681. https://doi.org/10.1056/nejmp2003539.

IMF. (2021, January 6). IMF Executive Board Concludes 2020 Article IV Consultation with the People's Republic of China. Retrieved February 11, 2021, from https://www.imf.org/en/News/Articles/2021/01/06/pr211-china-imf-executive-board-concludes-2020-article-iv-consultation.

IMF Vaccine Trade tracker. (2023). WTO. Available at: https://www.wto.org/english/tratop_e/covid19_e/vaccine_trade_tracker_e.htm (Accessed: 22 May 2023).

Imran, A., Posokhova, I., Qureshi, H. N., Masood, U., Riaz, M. S., Ali, K., John, C. N., Hussain, M. I., and Nabeel, M. (2020). AI4COVID-19: AI enabled preliminary diagnosis for COVID-19 from cough samples via an app. Inform Med Unlocked. 2020;20:100378. doi: 10.1016/j.imu.2020.100378. Epub 2020 Jun 26. PMID: 32839734; PMCID: PMC7318970.

International Monetary Fund. (2020, June 24). World Economic Outlook Update, June 2020: A Crisis Like No Other, An Uncertain Recovery. IMF. http Jiankangjie. (2020, November 23). 9 AI assisted medical device obtained Level III certificates from NMPA. Retrieved January 29, 2021, from https://www.cn-healthcare.com/article/20201123/content-546502. htmls://www.imf.org/en/Publications/WEO/Issues/2020/06/24/WEOUpdateJune2020.

Istepanian, R. S. H., and Woodward, B. (2016). m-Health: Fundamentals and Applications (IEEE Press Series on Biomedical Engineering) (1st ed.). New Jersey , U.S.: Wiley-IEEE Press.

Johns Hopkins University. (2020, March 12). Mortality Analyses. Johns Hopkins Coronavirus Resource Center. https://coronavirus.jhu.edu/data/mortality.

Johnson and Johnson, JLabs. (2020, May 12). Challenges & Opportunities Post COVID-19 — Mandarin. JLABS. https://jlabs.jnjinnovation.com/videos/challenges-opportunites-post-covid-19-mandarin.

Khan, K. (2020). Tracking the coronavirus pandemic with ai: BlueDot featured on 60 Minutes, Department of Medicine. Available at: https://deptmedicine.utoronto.ca/news/tracking-coronavirus-pandemic-ai-bluedot-featured-60-minutes (Accessed: 22 May 2023).

Lazarus, J. V., Wyka, K., White, T. M., Picchio, C. A., Gostin, L. O., Larson, H. J., Rabin, K., Ratzan, S. C., Kamarulzaman, A., and El-Mohandes, A. (2023). A survey of COVID-19 vaccine acceptance across 23 countries in 2022. Nat. Med. 2023 Feb; 29(2): 366–375. doi: 10.1038/s41591-022-02185-4. Epub 2023 Jan 9. PMID: 36624316.

Leonard, E., de Kock, I., and Bam, W. (2020, June). Investigating the barriers and facilitators to implementing an eHealth innovation into a resource-constrained setting: A South African case study. In 2020 IEEE International Conference on Engineering, Technology and Innovation (ICE/ITMC) (pp. 1–7). IEEE.

Mallapaty, S. (2021, October 14). China's Covid vaccines have been crucial - now immunity is waning. Nature News. Retrieved May 16, 2022, from https://www.nature.com/articles/d41586-021-02796-w.

Mars, M. (2013). Telemedicine and Advances in Urban and Rural Healthcare Delivery in Africa. Progress in Cardiovascular Diseases, 56(3): 326–335. https://doi.org/10.1016/j.pcad.2013.10.006.

McMorrow, R., and Leng, C. (2022, June 28). "Digital handcuffs": China's Covid Health Apps govern life but are ripe for abuse. Subscribe to read|Financial Times. https://www.ft.com/content/dee6bcc6-3fc5-4edc-814d-46dc73e67c7e.

Medica Presentation on telehealth solution in Germany (2019). Barmer.

MENG Qingyue, China Center for Health Development Studies, Peking University YANG Hongwei, and China Nat. (2015). People's Republic of China Health System Review (7th ed., Vol. 5) [E-book]. World Health Organization. https://apps.who.int/iris/bitstream/handle/10665/208229/9789290617280_eng.pdf?sequence=1&isAllowed=y.

Ministry of Foreign Affairs - Embassy of the People's Republic of China to the Kingdom of Netherlands. (2021, January 28). Notice on Chinese Citizens in the Netherlands cannot take transfer flights to Beijing. WeChat Public Account. https://mp.weixin.qq.com/s/udKdonDnZtj-qrFaZ-SiJA.

Murray, C. J. L. (2012, June 16). A framework for assessing the performance of health systems/C. J. L. Murray and J. Frenk. Retrieved January 1, 2021, from https://apps.who.int/iris/handle/10665/57320.

National Medical Products Administration. (2019, July 3). Summary of Review Standard for Deep Learning Assisted Medical Device Software. Retrieved January 29, 2021, from https://www.cmde.org.cn/CL0030/19342.html.

Nations must cooperate to offset economic fallout of covid-19, ex-PBOC chief says. South China Morning Post. (2020, May 9). Retrieved May 16, 2022, from https://www.scmp.com/economy/china-economy/article/3083644/coronavirus-nations-must-work-together-support-global-economy.

Nature Publishing Group. (2022, February 9). Africa is bringing vaccine manufacturing home. Nature News. https://www.nature.com/articles/d41586-022-00335-9.

Neville, S., and Cocco, F. (2022, August 30). The growing evidence that covid-19 is leaving people sicker. Financial Times. https://www.ft.com/content/26e0731f-15c4-4f5a-b2dc-fd8591a02aec.

OECD. (2022). Chapter 2. The health impact of COVID-19. Retrieved [Feb 28th 2022], from https://www.oecd-ilibrary.org/sites/b0118fae-en/index.html?itemId=/content/component/b0118fae-en.

Person, and Jones, M. (2022, May 3). Snarled-up ports point to worsening global supply chain woes - report. Reuters. Retrieved May 16, 2022, from https://www.reuters.com/business/snarled-up-ports-point-worsening-global-supply-chain-woes-report-2022-05-03/.

Person. (2022, March 18). Chinese officials urge elderly to get COVID vaccine, cite lesson of Hong Kong. Reuters. Retrieved May 16, 2022, from https://www.reuters.com/business/healthcare-pharmaceuticals/covid-vaccination-rate-people-aged-over-80-china-relatively-low-official-2022-03-18/.

Philips. (2020, May 19). Philips telehealth solution supports self-isolating COVID-19 patients in Australia. https://www.philips.com/a-w/about/news/archive/standard/news/articles/2020/20200428-philips-telehealth-solution-supports-self-isolating-covid-19-patients-in-australia.html.

Polonsky, W. H., and Fortmann, A. L. (2020). Impact of real-time continuous glucose monitoring data sharing on quality of life and health outcomes in adults with type 1 Diabetes. Diabetes Technology & Therapeutics, 23(4): 1–8. https://doi.org/10.1089/dia.2020.0466.

Project Overview' Safe paths (2020) MIT Media Lab. Available at: https://www.media.mit.edu/projects/safepaths/overview/ (Accessed: 22 May 2023).

Siemens. (2020, July). New Criteria for a New, Smart Building Era. Simens Smart Building White Paper. https://new.siemens.com/global/en/products/buildings/contact/smart-building-whitepaper.html.

Singh, D., Kumar, V., and Vaishali, and Kaur, M. (2020). Classification of COVID-19 patients from chest CT images using multi-objective differential evolution-based convolutional neural networks. Eur. J. Clin. Microbiol. Infect. Dis. 2020 Jul; 39(7): 1379–1389. doi: 10.1007/s10096-020-03901-z. Epub 2020 Apr 27. PMID: 32337662; PMCID: PMC7183816.

Soares, F. A novel specific artificial intelligence-based method to identify COVID-19 cases using simple blood exams. MedRxiv (2020).

South Africa bolsters its COVID-19 response. (2022). World Health Organization. World Health Organization. Available at: https://www.afro.who.int/health-topics/coronavirus-covid-19/south-africa-bolsters-covid-19-response (Accessed: January 2, 2023).

Student, J. (2020, November 25). AI puts Moderna within striking distance of beating COVID-19. Retrieved January 22, 2021, from https://digital.hbs.edu/artificial-intelligence-machine-learning/ai-puts-moderna-within-striking-distance-of-beating-covid-19/.

STUDY BASED ON LOCAL DATA REAFFIRMS THAT MRNA VACCINES OFFER BETTER PROTECTION AGAINST COVID-19. Singapore Ministry of Health. (n.d.). Retrieved May 16, 2022, from https://www.moh.gov.sg/news-highlights/details/study-based-on-local-data-reaffirms-that-mrna-vaccines-offer-better-protection-against-covid-19.

Tang, F. (2020, May 9). Nations must cooperate to offset economic fallout of covid-19, ex-PBOC chief says. South China Morning Post. Retrieved May 16, 2022, from https://www.scmp.com/economy/china-economy/article/3083644/coronavirus-nations-must-work-together-support-global-economy.

The Federal Reserve. (2020, December). Summary of Economic Projections. Retrieved from https://www.federalreserve.gov/monetarypolicy/files/fomcprojtabl20201216.pdf.

The World Bank: Covid-19 vaccine deployment tracker. COVID-19 VACCINE DEPLOYMENT TRACKER. (n.d.). Retrieved May 16, 2022, from https://covid19vaccinedeploymenttracker.worldbank.org/tracker/tables.

The World Bank. (2020, July 15). Martin Wolf CBE. World Bank Live. https://live.worldbank.org/experts/martin-wolf-cbe.

University of Sydney. (2020, March 27). COVID-19: the genetic quest to understand the virus. Retrieved January 25, 2021, from https://www.sydney.edu.au/news-opinion/news/2020/03/27/genetic-quest-to-understand-covid-19-coronavirus.html.

van Ginneken, B. (2020). The Potential of Artificial Intelligence to Analyze Chest Radiographs for Signs of COVID-19 Pneumonia. Radiology, 204238. https://doi.org/10.1148/radiol.2020204238.

Varshney, U. (2014). A model for improving quality of decisions in mobile health. Decision Support Systems, 62: 66–77. https://doi.org/10.1016/j.dss.2014.03.005.

Verity, R., Okell, L. C., Dorigatti, I., Winskill, P., Whittaker, C., Imai, N., Cuomo-Dannenburg, G., Thompson, H., Walker, P. G. T., Fu, H., Dighe, A., Griffin, J. T., Baguelin, M., Bhatia, S., Boonyasiri, A., Cori, A., Cucunubá, Z., FitzJohn, R., Gaythorpe, K., Green, W., Hamlet, A., Hinsley, W., Laydon, D., Nedjati-Gilani, G., Riley, S., van Elsland, S., Volz, E., Wang, H., Wang, Y., Xi, X., Donnelly, C. A., Ghani, A. C., and Ferguson, N. M. (2020). Estimates of the severity of coronavirus disease 2019: a model-based analysis. Lancet Infect Dis. 2020 Jun; 20(6): 669–677. doi: 10.1016/S1473-3099(20)30243-7. Epub 2020 Mar 30. Erratum in: Lancet Infect Dis. 2020 Apr 15: Erratum in: Lancet Infect Dis. 2020 May 4: PMID: 32240634; PMCID: PMC7158570.

World Economic Forum. (2019, May 8). These are 7 of the most exciting breakthroughs in healthcare today. https://www.weforum.org/agenda/2019/05/healthcare-technology-precision-medicine-breakthroughs.

World Health Organization. (2022). Covid-19 vaccines. World Health Organization. Retrieved May 16, 2022, from https://www.who.int/emergencies/diseases/novel-coronavirus-2019/covid-19-vaccines.

World Health Organization, World Meteorological Organization, & United Nations Environment Programme. (2003). Climate Change and Human Health (pp 16–17 ed.). World Health Organization.

Wu, C., Chen, X., Cai, Y., Xia, J., Zhou, X., Xu, S., Huang, H., Zhang, L., Zhou, X., Du, C., Zhang, Y., Song, J., Wang, S., Chao, Y., Yang, Z., Xu, J., Zhou, X., Chen, D., Xiong, W., Xu, L., Zhou, F., Jiang, J., Bai, C., Zheng, J., and Song, Y. (2020). Risk Factors Associated With Acute Respiratory Distress Syndrome and Death in Patients With Coronavirus Disease 2019 Pneumonia in Wuhan, China. JAMA Intern Med. 2020 Jul 1;180(7):934-943. doi: 10.1001/jamainternmed.2020.0994. Erratum in: JAMA Intern Med. 2020 Jul 1; 180(7): 1031. PMID: 32167524; PMCID: PMC7070509.

Xiamen Airlines. (2022, April 26). Notice on covid-19 nucleic acid test and serum antibody test for passengers taking flights to China (updated on April 26, 2022). Retrieved May 16, 2022, from https://www.xiamenair.com/brandnew_EN/about/business-notice/notice2022042456448.html.

Xiao, X., Califf, C. B., Sarker, S., and Sarker, S. (2013). ICT Innovation in Emerging Economies: A Review of the Existing Literature and a Framework for Future Research. Journal of Information Technology, 28(4): 264–278. https://doi.org/10.1057/jit.2013.20.

Xue, K. (2022, April 22). New details of Shanghai Nursing Home Covid deaths suggest city is overwhelmed. The Wall Street Journal. Retrieved May 16, 2022, from https://www.wsj.com/articles/shanghai-nursing.

Appendix

Ali Health Interview Notes

1. **The major businesses for Alihealth currently originates from health product sales and online business platforms, medication origination trace services, smart medical and personal health management services. How does Ali Health balance the input and output of the three business sectors to establish a sustainable business model?**

    1. Currently, the main user groups for three different platforms vary, they include patients, healthcare givers (doctors and nurses). hospital managers, so it is difficult to compare user behavior and business model of three business segments.

    2. Ali offers health care services in Alipay app and Taobao app, and has a large user base. It is easier to develop B2C businesses. (Business to Customer business).

    3. Currently, profits come from online drug sales. This is because in the short term, it is difficult to transfer users from offline to online to seek healthcare help. Now Ali's advantages lie in the fact that it has a large product platform and provides different types of services, such as free healthcare consulting services, so it is possible to attract users from free healthcare consulting and lead users to pay for other types of services, such as purchasing medicines online.

2. **Which types of wearable devices can be collected to Alihealth app? How is it being used now?**

    First of all, there are very few people using wearable devices to continuously monitor health related conditions, with only about 1% managed to do so. Users are not accustomed to continuously collecting health related data.

    Secondly, it is not an industry wide approach to integrate all the healthcare related data of users into a single platform. There is no industrial agglomeration effect. At the same time, the ownership right of the user's personal data is not clear. This resulted in the fact that data collected from personal medical devices cannot be shared on third-party platforms. For example, data collected from Omron Glucose Meter cannot be shared with third-party platforms, so it is difficult to form systematic personal medical data. The current business environment is not particularly supportive for user bio-metric information sharing.

Thirdly, the data generated from wearables are scattered, doctors cannot tell whether the data collected by users is reliable or not. If there is a gap between the data collected in hospitals and data from personal devices, doctors are more willing to believe in the data obtained in hospitals.

To sum up, due to insufficient medical information and unsupported business environment, doctors cannot use the data collected from individual wearables. The attempt to use the data collected from wearable devices or medical devices at home has just started. Wearables user experience is easier to improve, with the difficulty often lies in changing users habits.

3. **Currently, smart healthcare services, such as medical imaging processing, outpatient management, improving interoperability of medical institutions, telemedicine, internet-based hospitals, AI assistant for diagnosis and medical resource are offered by services provided by Alibaba. How does Alibaba serve the needs for different levels of healthcare givers? Does the current business model of Alibaba help to facilitate the current medical reforms in China?**

It is hard to tell, because different medical institutions have different demands for various healthcare scenarios.

The reform to establish a gate keeper in the healthcare system has failed, at least it is going on a slow pace. Compared with five years ago, probably only 1 more percent of patients have used the graded healthcare system. Alihealth has not tried to build the platform to facilitate the establishment of the gate keeper in the healthcare system. The main reason is that most patients would flow from lower level hospitals to higher level hospitals, rather than the other way around.

There are several issues embedded in community healthcare centers and township care centers or village clinics. Primary level of healthcare facilities do not have enough equipment or medication choices for patients with more serious disease or symptoms. More importantly, the training system for GPs (5 years of education + 3 years of training at hospitals) just started a few years ago. China does not have enough GPs to cover demand from patients.

Tencent is trying to connect doctors but there are very few tencent clinics (only in Beijing, Shen Zhen and Chengdu). Ding Xiang Yuan is trying to set up private clinics, where patients need to pay higher out-of-pocket ratios. Therefore, this is against the family doctor or GP notion.

The market for patient registration online is half-full. Almost half of all patients are getting used to make an appointment in advance and get registered online for a session with a doctor.

The nature of the telemedicine is to transfer information. If it is possible to communicate in advance via wechat, then it is possible to perform clinical guidance and medical interventions with the help of tele-communication methods.

Alihealth is developing medical services targeted at various scenarios with AI. It totally depends on how doctors plan to use these tools. For instance, the AI assistant for voice input will be more useful when doctors need to write down a large amount of patient records. Edge computing is more useful for updating algorithms in cardiac pacemakers.

Demand of primary level healthcare clinics lies in AI assisted diagnosis system and AI assistant for making prescriptions. Doctors are getting used to AI assistants. They are willing to use AI guide for doctor registration services or hospital bed needs projection system based on algorithms analysis. In Zhejiang province for instance, almost 80 percent of primary level hospitals are now equipped with AI assisted medical image recognition system.

4. **How has the smart health solutions provided by Alihealth is helping to realize value based healthcare, and promoted the patient centered, data-based decision making process? Has the smart health solutions improved patient satisfaction, lower medical cost, promoted diagnosis efficiency and precision, and helped to avoid doctors' burning out?**

Alihealth currently provides service packages on a half-consultancy, half solution packages basis. The services provided is on a highly personalized basis. Different hospitals usually have very different requests. Hospitals do not get standardized solutions, with only 70 to 80 percent of the services provided are standardized. Now there is a strong market demand for the service packages offered. This is due to the fact that there are seldom any consulting firms offering standardized services, and there is no consulting firm focusing on healthcare services only. At the same time, some management teams of the hospitals are not willing to accept the service packages.

For instance, Alihealth is able to offer services which can improve two KPIs significantly—the standardization level of the patient records and reduction in patient infection rate. We were able to improve standardization level of the patient records from 20–30 percent to 50–60 percent; we can also lower patient infection rate by about 30 percent. However, there is no clear standards for value evaluation for the services we offered. We do not know if it is because the management team for the hospital paid attention for these indicators, or it is because our solution packages have worked.

Meanwhile, we offer systematic solutions for certain points of the services provided to hospitals.

Currently we only have sporadic and unsystematic solutions for NCD management and aging population.

We offer services to local governments and the public finance system pays the cost.

5. **What is the acceptance level of smart health solutions with patients, doctors and hospitals?**

Doctors are not threatened by smart health solutions. They are getting used to it, with the hospital management team willing to use such solutions.

6. **Alihealth is currently offering the smart care service package, aiming to target the entire health continuum; for instance, the health secretary to automatically answer questions about healthcare related issues. Are doctors willing to sacrifice part of the privacy to upload their data. What are the acceptance level for different types of monitoring technology?**

Currently, there are very few users which consult doctors online. Mainly users complain about sex related issues (for instance, andrology or gynecology related issues) and other issues which may cause embarrassment in face-to-face interactions. Meanwhile, online consultation are mainly about offering advices and recommendations, it belongs to consultancy related services. Alihealth does not offer electronic prescriptions. All the services are paid out of pocket by patients; the services may not suffice for refund from the basic medical insurances. The main argument is to get users to use medical services online, and to attract users to use Alihealth services. For instance, users who get accustomed to use free consultancy services online may buy services such as medical examination or losing weight services.

There are several barriers for establishing personal portable EHR system.

First of all, data does not belong to patients in China. Instead, data controllers such as doctors who perform medical examinations, medical device producers, hospitals and regulatory agencies all have the power to deal with data as they wish. Because of this, rarely hospitals would provide electronic medical records for patients. If patients demand such information, there might be hospitals which provide such information. However, patients seldom demand such all of their electronic health records. Most likely patients would demand CT images or test reports. Possibly the hospitals at their home city do not have such equipment to perform those tests. In this case, they get tests done in big hospitals in big cities, and then show the reports to doctors at home hospitals.

Secondly, different hospitals and different doctors have various standards for interpreting test reports. Patients can get examined at Peking Union Medical College Hospital with the results not recognized by 301 Military Hospital (another top-tier university hospital in China). In the meantime, doctors may order a test in different contexts, usually based on their own experiences. After a few months, patient's condition may evolve. Because of this, public EHR system is difficult to establish.

Thirdly, allergies of patients are private information of the patient. It only shows on the summary report when patient leave the hospital. Because of regulation, and patient complaints, it is difficult to upload such information to the internet.

Fourthly, whether users for medical services shall share information or not is contentious. Not all related stakeholders are willing to share such information.

Fifthly, the medical regulation system in China is a top-to-bottom system. There might be data accuracy issue in the top-to-bottom system. In comparison with healthcare associated infection rate of about 3 percent in the EU, U.S. and South-East Asia countries, China maintains a rate of about 0.5–1%. There are cases of concealing and mis-representation for data reporting. Data about chronic disease can be an over-statement of reality as well.

7. **How does Ali solve the problem of data isolation in hospitals? How is the population health management business doing now? Who are the main customers? How to balance various stakeholders in the population health management business? Which stakeholder is the main driver?**

Some local governments put a lot of emphasis on data integration in smart city. If the head of the local government or the local party's leader decided for a project, then the public finance would pay for such services. By only relying on the local municipal health commission, it is difficult to break the current chain of interests. For instance, the city of Hai Kou now plans to use smart city brain to process all the data about transportation, tourism, medical services, IoT, mobile office. Hai Kou plans to implement smart city brain solution to improve the operation efficiency and capacity of the city. The current project involves computing platform, data resources platform, security platform.

Alihealth offers services in data integration and processing, and improve the usability of data with AI. The core problem is not about technology, but about data operation.

Take an example of allergy condition of the patient, the patients' records might exist in doctors' treatment records or nurses' logs. The data can be quite scattered from different sources. Alihealth uploads all the related records from doctors and nurses to the cloud, and seize more sensitive information in the cloud platform.

The B2B services are mainly offered for governments or hospitals and other institutional users. The services are mainly about data processing platform. We can promote the electrification of the hospital information system of the hospital from 20–30 % to 60%.

8. **How long would it take for algorithms for image recognition, voice recognition, OCR to apply in medical related scenarios.**

Currently, the capital + technology model is very mature. It takes about 3 months to finish the design of the UI. It is mainly about the users' needs and scenarios. It is difficult to identify the core needs of users in medical related scenarios. For instance, voice recognition is not the core needs in telemedicine.

The needs for primary level of healthcare facilities is to reduce cost. Therefore, 80% of all healthcare facilities in Zhe Jiang Province are equipped with AI assistants for image processing. At the same time, the utilization rates for AI assistants for prescribing and diagnosis platform is relatively high.

Interview Records of the Chinese Center for Disease Control and Prevention in China

1. **What is the role of primary level healthcare institutions (community healthcare centers, village clinics, county-wide hospitals) in the healthcare continuum? How to strengthen the role of primary level healthcare institutions in the ongoing reform for the healthcare industry? How to alleviate rising costs for healthcare services and improve access to health care services for the Chinese population.**

The grading of the healthcare institutions played a role in history for instance, barefoot doctors. The system has its advantages.

For chronic disease management, primary level of healthcare facilities shall play the main role with the large amount of patients with COPD.

The low utilization rate is due to the fact that patients do not trust primary level of healthcare facilities. Patients do not recognize the value of primary level of care either. In the meantime, the nurturing of good habits for seeking medical services and to improve the security of public health policy. The UEBMI, URBMI (now the urban and rural residents basic medical insurance) can improve the utilization rates of primary level healthcare facilities and leverage the use between third-tier hospitals and primary levels of hospitals.

The primary level care facilities are tackling the shortage of medical equipment and the lack of medicine supply.

2. **What is your attitude towards smart health solutions and its future application in China?**

Utilization of smart health solutions can be related to the technology matureness level of the tools, how to use tools and the recognition level of the tools.

Now with the advancement of technology, the sensitivity of wearables is improved. In the meantime, the role played by AI still yet to be seen. Whether policy makers shall use AI and wearables as an assistant, or to replace human beings is upon discussion. China does not have the entry standards for wearables and AI use in the medical industry. The cautious attitudes towards use of AI in the medical industry should be taken.

Family doctor and GPs should effectively act as gate keeper in the healthcare system. We are currently improving the role of family doctors played in the healthcare industry. Relevant efforts include reforming the GP education system in China, implementing the 5+3 model (5 years of clinical courses + 3 years of standardized clinical training) and the improve wages for the family doctors. To transform the system where the family doctors acting as gate keepers can change the habits of patients accessing medical help. It can effectively help to resolve the chaotic situation in medical access. The interviewee has a positive attitudes .

3. **What is the role played by EHR in the healthcare system in China? How to establish the EHR which can be synchronized with HIS and patients' home medical devices? How to make sure EHR interoperable in different healthcare institutions.**

Some regions, such as Beijing has implemented the solution of 'Jing Yi Tong', which allows patients to carry one card and use it in some third-tier hospitals. Meanwhile, third-tier hospitals can recognize the test results from one another, and allow doctors to access patients' allergy history, family medical history and operation records.

4. **What is the current application and cases of big data in population health management and disease prevention? Which parts of China has done a better job to monitor population health with big data?**

Yi Chang – Hubei Province, Ning Bo-Zhe Jiang Province, Shanghai and Beijing has a more thorough health related data collection system. Different health issue regulation authorities departments has made it possible medical data to be shared and inter-operable. In the meantime, it is possible make decisions based on data projection. It is possible in some communities to monitor patients with NCD.

The National Health Commission has a migrant population service center to manage healthcare related issues for groups such as farmers who lost their land and flow in to the city.

Realizing data interoperability should be realized step by step, it will start with experimental programs in certain regions and then the system can be promoted nationwide.

5. **How to realize the value-based healthcare system and the included integrated healthcare service, big data in healthcare, and health education and health promotion in China? Currently, what is the role played by preventive healthcare and outpatient management services in chronic disease management?**

Value based healthcare programs matter, because it will make healthcare process more standardized, find better solutions and is the trend for the future.

Chronic disease management has received more and more attention recently. Now the chronic disease prevention program incorporates early detection, diagnosis and treatment program, vaccination, free health examination programs for the elderly.

It is true there is lack of investment in preventive healthcare services, only about a few percent per year of all healthcare expenditures in China. (I CANNOT FIND PUBLIC DATA to validate mostly because there is no data on how much does the government spent on preventive healthcare, or there is no budget specially devoted to preventive healthcare in China; checked WHO, World Bank, National Statistics Bureau).

6. **Is there any practice in China to use IoT devices to collect data to improve patient satisfaction level, improve diagnosis quality and to control medical costs from rising in China?**

The implementation of Internet of Healthcare solutions can be seen from technological standards and legal regulations and several other perspectives.

It takes some time from the technology becomes available and its implementation. In the meantime, the smart healthcare solutions have its advantages and disadvantages. It may bring medical accidents.

The implementation of IoT in the healthcare industry needs to be taken with caution. There is a limit amount of role which can be played by technology and it cannot replace human beings. In some respects, such as preventing elderly from falling at home or to realize tele-medicine, it can be very useful.

7. **What are the prospects of using IoT and big data in disease prevention, diagnosis, treatment and home care?**

In CCDC, we mainly used big data for academic research. Currently, the analytical results have been applied in policy making. Currently, the implementation process shall start with some experimental programs.

Data collection for NCDs starts with empirical research from top to bottom, sample taking, and check for missing and mispresented data. Also, there exists data reporting from bottom to top.

For communicable disease management, CCDC use a real-time monitoring system. NCD data collection comes by month, season or year. It can take 1–2 years for the data to become publicly available. Because it requires a lot of work to verify the data and check for missing data. In the meantime, the data is summarized by each department in CCDC annually.

8. **Can you please share practices to promote healthy living style with IoT devices via controlling diets, exercises to prevent and manage chronic diseases (hypertension, cardiovascular disease, obesity, and asthma et al.)?**

It depends on specific cases of usage. Technology shall not intervene and change the original life style of the population. Instead it should nudge users to change bit by bit.

Chronic disease shall be prevented via controlling diets, exercises, smoking, alcohol control and vaccination. Early detection of disease also should be emphasized. Risk factors leading to disease shall be identified.

Community healthcare centers should play a more important role in chronic disease prevention. CCDC is responsible for diverting the population to healthy living style.

It is difficult to ask the medical insurance funds to pay for the IoT solution packages. Because the basic medical insurance schemes (UEBMI, URBMI, the NCMS, the latter has been integrated since 2016) usually cover inpatient costs. In the future, private insurances may cover the cost of the devices.

The control and manage NCDs via smart healthcare solutions shall be gradually implemented.

The leading NCDs in China ranked by incidence rates are hypertension, diabetes, cardiovascular diseases, and cancer. The leading NCDs ranked by death causes are cardiovascular disease (top 1 death rate) and cancer (top 2 death rate).

Interview Records Zhongshan Hospital China - Smart Health Solutions for Chronic Disease Management

Do you think that the current use of wearable devices, watches, bracelets, etc., is helpful in preventing chronic diseases (cardiovascular disease, diabetes, high blood pressure, osteoporosis, etc.)? For example, the use of wearable devices for monitoring user›s exercise data, diet data, and the like.

The so-called wearable devices have the advantage in convenience of use and data monitoring continuity.

If so, do you think the medical community accepts the data obtained from wearables?

The premise for use of data obtained from wearables, except for the convenience provided, is data accuracy. The most important aspect for medical devices is accuracy. To monitor heart rate—slow or fast, or during sports sessions is a function also used by healthy population. For patients who are sick, data accuracy is more important than convenience of monitoring. If there is a method to monitor data both accurately and conveniently, it will be the best.

Currently, use wearable devices in medical settings need to follow the approval of medical devices. Is it necessary to set the technical standards for wearable devices use in medical settings separately?

It is definitely necessary, of course, it is necessary. For the accurate and convenient use of wearable devices in the medical setting, separate technical standards for wearable devices must be available.

For example, patients with hypertension can obtain the two-point based ECG from the Apple Watch 4, or the changes of blood pressure in the last month from the family blood pressure monitor; do you think you will consider this type of data?

I would like to consider such data and take it as a reference when I make my diagnosis. Whether the devices can provide accurate data monitoring however demands evidence based studies. There is a need to compare with the traditional/classical way of monitoring. If the data about heart rate or blood pressure can be comparable in accuracy with the traditional way of monitoring, then doctors may take the data into consideration. If the data is highly accurate, then doctors may use the data to make further diagnosis. If the data is not accurate enough, then it makes no sense for doctors to use it. Doctors need to follow the national standards.

Personally, I think ECG is more useful than the mere heart rate readings.

Now wearables are able to monitor VO2 and heart rates, and also provide nutrition monitoring; for example, smart phones with cameras can automatically analyze nutrition contents for the elderly. What do you think of the value of such functions?

It is good for patients to have such functions. For doctors at cardiovascular department, what we care most is the blood pressure and heart rate changes. The cardiovascular department cares less about nutrition of the outpatients than the diabetes department physicians do. Of course we care about blood cholesterol level and diets with low sugar and low salt contents.

Do you think the medical devices to use in home settings, such as Bluetooth equipped blood pressure monitor, blood sugar monitor, scale are helpful for prevention of chronic diseases?

From our perspective, the data from medical devices at home settings is very valuable. The cardiovascular department does not cover diabetes, but we care about hypertension. The blood pressure changes at the hospital tests do not embed significant meanings. However, the self-tests from home can be meaningful with the major guidelines recommend blood pressure monitoring from self-test at home. Sometimes physicians do ask patients to come to the hospital to perform the 24 hour continuous blood pressure monitoring. It is a lot of trouble for the patients, which requires making appointments, and take the device down at the hospital. This makes the blood pressure monitoring at home more meaningful.

If it is helpful, do you think the medical community will accept the data acquired from the medical devices at home?

The data from home medical devices are not connected to the EHR system; the lack of interoperability made it difficult for doctors to accept data from the telehealth devices including medical devices at home.

If you are helpful, do you think the medical community currently accepts data from home medical devices?

But the accuracy of the problem detection, which is the most critical, is that its data is definitely valuable, just to say how accurate you can be.

Now the country only accepts the oscillometric method to monitor blood pressure, and then only accepts the blood pressure data measured by the pressurized pump. Many users may not need to be so precise, because if he uses a pressure pump to measure blood pressure, the 24-hour monitoring may not be necessary, because the patient may or sleep and normal life will be affected.

Just like we say 24 hour ambulatory blood pressure monitoring. Because I recently gave my husband a 24-hour ambulatory blood pressure monitor, he told me that he was so sad that he had to take a breath every hour and he had to take a breath every hour. He felt that he was often interference.

Omron offers a Generation Zero. There is a PPG sensor, then that is may be measured by light pressure pump may not be accurate, but the trend of changes in blood pressure can be measured, but not through the C FDA audit.

Because the Food and Drug Administration feels that this type of monitoring is not accurate enough, because inaccuracy will affect medical behavior, that is, if the patient measures it high, is the doctor adding or not taking it? Inaccurate doctors cannot judge what to do next.

For patients who already have chronic diseases, do you think that a combination of wearable devices, home smart medical devices, and family doctors can help manage various chronic diseases?

A chronic disease is that the patient has been suffering from a long-term illness. The doctor at the top three hospital has given you the medicine, and it has been cured by the treatment plan. It is a chronic disease when you are basically stable. Community doctors often let patients take medication stare. Huashan Hospital has cooperation with community hospitals. Huashan Hospital has a chest pain center, which has a radiation surface. Below the patient will turn up, sometimes we will go down to the doctor for training. This is two-way. Sometimes Huashan Hospital feels that the patient can be discharged from the hospital, and the follow-up hospital will continue to manage the patient's condition. The question now is a great degree of patient choice, if the patient is not willing to accept lower hospital services, the latter may still turn back.

The idea of this kind of thing is very good, but there are many problems in practice. The wearable device is at least a few hundred yuan, and the patient may not be more than ten yuan in the hospital. Sometimes I will take the initiative to ask the patient, what should I do if the queue time is too long? However, the elderly feel that they have plenty of time. The cost of going to the hospital to see a doctor plus the transportation fee may add up to 30 yuan. Many older people are thinking about this. It may be necessary to wait until the current group of young people is old, and with the disease, the acceptance of this concept will be higher. But the current acceptance is not that high.

What are the main plans for hospitals to manage chronic diseases for patients? Have you developed a mobile phone a pp for managing user health, or for remote communication between doctors and patients?

Basically, the patient came to me on his own initiative. He was willing to come to me when he was willing to come to me. I didn't want me to go there. After the patient was discharged from the hospital, I basically did not communicate with the patient because there was no time. After the patient is discharged from the hospital, we will print a discharge summary for the patient. In the What kind of medicine, medication time to pay attention to what. We have a discharge recommendation, and then the nurse will give him some education outposts for health discharge before leaving the hospital. For example, if you want to do some tests regularly, you should do a small check on the discharge summary for a few months.

The main thing is that Chinese doctors are different from foreign doctors. I think you actually talk to a lot of Internet companies, including Tencent. The problem is that I don't have time to manage it. That is to say, we cannot rely on us to manage it. I can manage the doctors in the top three hospitals , but it is impossible to manage them in a meticulous manner. Family doctors in the community are not easy; I know doctors in several community hospitals, they are very busy. The main thing is that a doctor manages a lot of patients. There are too many people in China, but the number of doctors is not much. If the family doctor team of six people needs to manage a community of tens of thousands of people (take the Beijing Huairou Bridge Township community as an example), this is definitely not good, it is the management. When a doctor manages a patient population of one thousand people, it is impossible to see the data of each patient , so it is even less important to manage. When a doctor manages data for thousands of patients, probably I don't necessarily have time to look at it.

After the family doctor signed the contract, taking the Shanghai community as an example, it may be that the lonely old people in the same community could not move. They went to see it. Most of them are uncomfortable to come to me. Doctors have limited energy and the number of patients who can actively manage is very limited. Currently managing chronic patients, family doctors can come to the door, but there is no energy to carry out on-site service for all patients. Community doctors come to the door,if there is no additional income, there may be no enthusiasm to do it. If the country does not have enough input, the establishment of a family doctor contract can only be a slogan. So now the Internet medical service is to transfer offline medical behavior to the Internet. But the Internet is not good for some older people. It may take 10–20 years to wait until the people who accept Internet services are old, and these Internet medical applications may be better. But now in this situation, the elderly will not use Internet medical services, they can only go to see a doctor. At the same time, patients do not trust Internet medical services. Because many of my patients, that is, using an automatic sphygmomanometer to measure blood pressure, feel unreliable, must be measured by mercury blood pressure. Older patients still have no trust in wearable devices and Internet medical services.

Is there a data island in the hospital? What do you think about establishing a regional and national patient health data repository?

Hospital because of financial independence, so the administrative operation is relatively independent. At present, the hospital has no motivation to unify the medical record standards and allow electronic medical records to communicate. I can see the data in the hospital.

These are problems at the national level and are not problems that frontline doctors can solve.

What do you think about the role of electronic medical records in the current diagnosis and treatment process? Do you think that the electronic medical record has increased or reduced your workload?

When I was writing a medical record, there were fewer patients and slower turnover. Now the patient is more and the turnover is faster. Electronic medical records should always be faster than handwriting.

It would be best if these tools were able to free the doctor from the heavy work. If I just talk about the medical record, I will do it. No, I don't know if this is something that our little doctor can decide. For example, the history of artificial intelligence writing, the hospital agreed to use, I can use. The hospital does not agree to use it, and doctors cannot use it. I didn't have this kind of tool similar to the voice input engine of Keda Xunfei, but I think it is still inconvenient. Because sometimes I still make mistakes, I have to change the mistakes, it is better to type faster.

Do you think that the role and application prospects of artificial intelligence assistant tools, artificial intelligence assisted diagnosis engine, artificial intelligence medical image processing engine, artificial intelligence voice recording, artificial intelligence medical record processing engine and other artificial intelligence tools in the medical process?

Like ECG report will have an automatic call me, this report is to analyze the ECG data is automatically made. The electrocardiograph only has measured data that is accurate, such as how wide and how high, but the diagnosis is often inaccurate. So doctors hardly look at these diagnoses.

What obstacles do you think currently have to achieve population health management? What are the application prospects?

Currently CDC (C DC) database data, as doctors do not have permission to view, can only be reported. However, at present, we are not reporting chronic diseases. Only cerebral infarction, myocardial infarction and tumor are reported. This situation, like the new diagnosis of high blood pressure , will not be reported. The data of the chronic disease center for the total number of chronic diseases in the country should be obtained through a sample survey. CDC reports cardiovascular disease every year.

Do you think these data are worthwhile if the data collection on chronic diseases is delayed by 2–3 years ?

This is at least a trend. The key management in the community is blood sugar, and their blood sugar will look. From a realistic point of view, the incidence of high blood pressure and high blood fat is high now.

Do you think that smart health care solutions are currently helping the health management of an aging population? What other applications might be in the future? If you want to promote smart healthcare solutions, what problems might you encounter in the future?

About 20% of my patients—30% of people can regularly measure blood pressure and blood sugar, and regularly manage blood pressure and blood sugar in a long period. Because I am looking at an expert clinic, my patient's compliance will be relatively better. A normal outpatient clinic may not measure blood pressure in one patient. At the same time, whether it is Internet medical treatment, wearable devices or IoT medical platforms, the patient's consciousness is the most important. These tools, even if the doctor feels valuable, many patients may not be conscious of the use. Just like the patient we were discharged from, the doctor gave the patient a good medicine. Many patients may not have taken this medicine after half a year. Either stop it or reduce it. These precautions are explained in the discharge summary, but the patient will stop taking the medicine or change the medicine. Some patients feel that the medicine is annoying, and taking medicine has side effects. This problem also exists in Europe and America. Now many platforms or a pp do things to remind patients, but the problem is that patients ignore this reminder. Wisdom medical solutions are services in essence, the most important being the main body and direction of use. It is useful to push a patient a lot of things, but some people don't use it.

The people that this thing can be involved are relatively small. First, users who are willing to pay for wearable devices are first of all economically sound. Second, the willingness to use the Internet in this way , willing to use this platform to manage their own healthy users, their own compliance is better.

What do you think of the construction of a value health care system? What role should hospitals and doctors play in building a value health care system?

The last time I saw one, AI medical treatment did not actually succeed. However, from the current medical reform situation in China, the burden on doctors is very heavy, but there is no corresponding compensation. The doctor's medical treatment fee does not rise, and the treatment cannot be improved. In this case, the contradiction between the patient and the doctor cannot be alleviated. Because the results of medical treatment are often not just the reason for doctors. It is also related to the patient's living habits and environmental factors.

What data is worth collecting in the medical process?

If the government needs to make a regional, national system, then all the information is worth collecting. Because the patient's information will be scattered, in the records of the doctors, nurses and discharge summary.

What do you think of the construction of a graded medical system? Can the construction of a grading medical system reduce the burden on the top three hospitals and ease the relationship between doctors and patients?

China is like this, a lot of good grade before treatment is mandatory for many years, is that you have to look at from the community health service stations, but there is a fixed point, is that you can only go to that hospital. Looking not only to go to the hospital to see a higher level. It was cancelled many years ago. Shanghai all your hospital you want to go home can go. Then all patients are hoping to come to the top three hospitals. Patients with medicine, he might be willing to go to other hospitals, because it's easy dispensing. However, if the patient is seeing a doctor, because it is not a problem with the equipment, most of them include the level of the doctor, and there is still a gap, so the patient feels that he needs to go to the top three hospitals. And tertiary hospital registration fee unlike the United States, for so cheap it does not matter, queue line up on the patient feel better. Therefore, I feel that the graded diagnosis is actually not so good.

What is the opinion of the Internet hospital?

It can only be followed up. Because there are many problems in the initial diagnosis, there are problems with Internet hospitals. This is mainly security. As a doctor, there is no time to communicate remotely with the patient. Because there are many people who come to talk to us about this problem, do a variety of patient follow-up. Because this is a hot spot now, there is a lot of speculation. As a doctor, I hope that the platform can be done very simply. Second, there are actually security issues, and I think it is very likely that patients will be missed. Third, doctors don't have time. Now telemedicine exists between hospitals and hospitals. Before applying, you need to apply for a doctor. I need a doctor there. I am free. The last time I went to telemedicine, I had a heart failure patient. I could only look at him from afar . I couldn't even see him. All the doctors were holding information and told me. I think that in this case, I can only give a broad recommendation. I can't follow the patient who lives in my ward so well. Still a bit unreliable.

Interview Records for the National healthcare security administration

Part I: Difficult to access healthcare, expensive healthcare services, and smart healthcare services

1. **What are the measures and policy responses for the migrant workers and their needs for high-quality healthcare services? Because of fast urbanization in China, a lot of farmers has lost their land with a large amount of them flowing in cities. This has intensified lack of access to medical resources in urban areas of China.**

   Currently, there are different medical insurance schemes for different types of residents. China has achieved universal healthcare coverage (UHC), with the basic medical insurance schemes covering around 1.37 billion people, roughly above 95% of all population. The UHC has three vertical and horizontal layers in policy schemes. Three horizontal layers refer to the bottom layer of medical aid, main layer and top additional layer. The bottom layer of medical aid scheme consists of urban and rural medical aid, social charity donations. The main layers consist of urban employee basic medical insurance, urban resident basic medical insurance, and new rural cooperative medical insurance. The urban resident basic medical insurance, and new rural cooperative medical insurance will soon integrate into the urban and rural resident basic medical insurance. The health security insurance system in China will soon turn into a system with three vertical layers and two horizontal layers. The additional top layer is made up of critical illness insurance scheme and out of pocket paid commercial complementary medical insurance.

With rapid urbanization and the migrant population flowing in urban areas, the UHC has managed to cover growing healthcare needs from the migrant population. To meet the demands for the migrant population and their needs for accessing healthcare services, the national healthcare security administration has taken three steps. The first step is to realize municipal level of interoperability for basic medical insurances, the second step is to provincial level of interoperability and the third step is to implement national level basic medical insurance interoperability. The administration has started to implement national level offsite medical settlement services in 2016.

The services for site-off medical treatment mainly covers the healthcare needs for the following segments of the population.

The first category is the re-settlement elderly who also become permanent residents to the city where he or she is moving to. The second category is the elderly who live in the city who lived at another city (without permanent residence registration). The third category is for people work at different cities other than their registration. The fourth category is for patients who are referred to other hospitals and doctors at a different city. These four groups can use their medical insurance cards at designated hospitals located at cities other than their registration for inpatient treatment and get real-time reimbursed. In 2018, there are 1.8 million people who has used such services.

The system administration: patients need to file for registration with the local medical insurance administration authority, and inform of the authority about their intended destination of use of the health insurance cards. By registration, they can use the medical insurance card at designated institutions in cities other than their registration and get reimbursement in real-time. The registration is necessary because the system cannot support 1.37 billion people of using their insurance cards on a long distance basis in real time. The fund management administration wish to ensure the patients can be referred to institutions with qualifications and there are about 3.65 million people who have filed for registration.

Services: To improve access for healthcare services for the migrant population, the bureau is promoting online registration, registration via telephone, over the app. For migrant workers, their destination for work may not be stable. It will help them to access healthcare services by promoting services conveniences.

2. **What is the attitude of national healthcare services administration towards smart health solutions including online family doctors and offline clinics, and wearables and their monitoring of the population? Are there any plans to integrate family doctor services into the coverage of basic medical insurances?**

The family doctor system and the smart health solutions is in the early stage of development. The services offered by the public health security administration authority is to pay for patients getting treated at public hospitals. The health security administration sits at the end of healthcare continuum, and it mainly secures payment and settlement for healthcare costs. The National Health Security Administration in principle holds a reserved attitude (wait-and-see attitudes) towards new businesses and hope there are standards to measure the effectiveness and quality of services offered by smart health solutions, the amount of services and fee standards. These questions can be answered from practices. Only with a mature service system, the healthcare security fund can catch up with reimbursement.

The GP acting as gatekeeper system has received support from basic medical insurances from the beginning. Currently, the supply side of GP services is small (with 309,000 doctors in China, around 2.2 GPs per 10,000 people at the end of 2018),[3] with little recognition level from the society, and little trust from patients. Some cities offer good GP services, for instance, community healthcare centers in Shanghai, with about 100 Yuan support per person annually. The services need recognition from the insured personnel, with patients willing to go to the services.

3. **What is the policy and measures targeting at the establishment of (hierarchical medical system) and held healthcare services become more accessible and affordable?**

The reimbursement ratio for the insured (referring to public medical insurance schemes) for medical care in a community healthcare center of Beijing is over 90%. Patients over 65 years old do not need to pay a registration fee, thereby making all services almost free. The

---

[3] https://www.thepaper.cn/newsDetail_forward_3374431

community health centers in Beijing offer over 800 types of medical supply in their pharmacies.

Personally, I feel community healthcare centers have much more roles to play than treating patients. The system would not work if community healthcare centers only plays the role of treating patients and cure diseases. The main role of community healthcare centers is not treating patients, but instead focusing on caring for the insured. Community healthcare centers cannot compete with tier-3 hospitals from treating patients; only by differentiating from tier-3 hospitals, can community healthcare centers thrive. 30% of the patients attending tier-3 hospitals do not need the specialist services because they cannot find doctors to consult there. Tier-3 hospitals need to be responsible for patients' health conditions.

How to improve utilization rate for primary level of healthcare services? Is it possible for patients to upload their data online to improve communication with doctors?

Seeing a doctor is not as easy as uploading the blood pressure data online. Treating patients demand more communication. Internal physicians do more work than just uploading data online. People need care, with technology more advanced, care for people is less and less. The fact that it takes two hours of waiting to see a doctor and two minutes to treat the patients is sad for the doctor, sad for the society and sad for the patient.

How do you feel about using IoT devices to monitor patients over the long term and help them to control their chronic conditions? What are the institutional barriers for realizing such scenarios?

Certainly, it would be useful, and it would be indeed be very useful. The implementation however demands for proactive participation of doctors and high compliance level from the patients, accurate and objective data. In the current healthcare system, doctors don't have the time to do such things. Therefore, it is reasonable for community healthcare centers to provide such services.

4. **Currently personal health expenditures account for 29.3% of all health expenditures, while total healthcare expenditures account for 6.2% of all GDP. The Health China 2030 Initiative decides to lower the ratio to around 25%. Compared with about the 15 percent of personal healthcare spending in total health expenditures in the EU countries, how will public healthcare expenditure change in the future?**

Personally, I believe the healthcare expenditures accounting for 6.2% is not a small amount. The key issue lies in the spending structure. Currently insufficient health expenditure and waste both exists.

Healthcare expenditure per capita is about 600 dollars, while it is about 10,000 dollars in the U.S. and about 6000 dollars in the UK. 600 dollars per capita is not low considered the reality of China. Take a decomposition of the expenditures, the pooling for urban employee basic medical insurance has reached around 5000 RMB per capita, with urban and rural resident basic medical insurance reaching about 500 RMB per month. It is key to improve access for healthcare for urban and rural residents.

5. **Currently, the amount of outpatient treatment by tier-3 hospitals in China is on average 1.73 billion annually, the amount of outpatient treatment by tier-2 hospitals is about 1.27 billion, while the amount of tier-1 hospital outpatient treatment is about 0.22 billion. The hospital bed utilization rate for tier 3 hospital is 98.2%, for tier 2 hospital is 845 while it is about 57.5% for tier 1 hospitals. The amount of healthcare facilities and the amount of outpatient/inpatient treatments do not leverage. Is it possible to turn tier 1 hospital into elderly long-term care centers?**

To target the needs for the aging population, now different cities are exploring different options. In every district of Beijing, there are a few elderly long-term care facilities, some tier-1 hospital and community healthcare centers are turning into elderly long-term care centers these days.

The elderly is easily subject to fall, or amnesia where they lose memory of their location. Is it possible to use IoT devices to help elderly to avoid such scenarios? Is there any chance that the basic medical insurance schemes are going to cover these costs?

Theoretically speaking, there is implementation potential for such devices. The most important factor here is figuring out how to pay for the devices. The basic medical insurance schemes will not be able to cover these costs. The IoT devices should be paid by whoever is using the devices. The basic medical insurance schemes cannot meet the needs for patients to get properly treated, and therefore cannot cover the cost of IoT devices.

Is it possible to rely on commercial insurances to pay for IoT devices? With many high-end elderly care institutions offer door-to-door visits, will ordinary long-term care facilities offer the same services?

Wearable devices and other IoT devices cost will not be covered by basic medical insurance schemes lately.

How much does preventive healthcare costs account for in total healthcare expenditures?

The basic healthcare insurance schemes in principle does not cover preventive healthcare costs. Now primary level healthcare facilities offer basic healthcare services, which covers setting up health records, health education, vaccination and healthcare advices for new-born and pregnant women, elderly care, chronic disease management, severe mental disease obstructed patient care, etc. These public healthcare functions belong to chronic disease management, health management, health consultation, etc. Average subsidy paid by public finance system is about 50 RMB, where 45 RMB is used for prevention and health management functions. Considering the amount of population in China, it is not a small amount for the public finance system in China.

6. **In 2015, about 3.6% of all patients get reimbursed via commercial insurance schemes. Take the decomposition for personal healthcare expenditures in China of 2014, cash payment accounts for 72.4% while reimbursement with commercial insurance schemes account for about 10.2%. Chinese population are not used to pay for commercial insurance premiums and get healthcare costs reimbursed. Is there any plans for the National Healthcare Security Administration to combine the basic medical health insurance schemes and commercial insurance schemes to lower the ratio of personal expenditures in total healthcare costs?**

In principle, the commercial insurance schemes have nothing to do with the basic healthcare insurance schemes. To combine both insurance schemes is wishful thinking.

The core of the problem is Chinese people have no money to pay for the commercial insurances, and secondly, private insurance companies have not developed appropriate products which meet the population needs. (Most private insurance schemes have a quota for reimbursement, and the coverage of such schemes overlaps with basic medical insurance schemes.)

The insurance companies in China are in development stage, with private insurances cannot take such responsibilities. Therefore, the basic healthcare insurance schemes in China are getting bigger in scale. With the public healthcare insurance schemes getting stronger, the private insurance schemes lose market share. (Jokingly saying, the healthcare expenditures per capita for Chinese is about 600 dollars. Comparing to health expenditures for US citizens sitting at 10,000 dollars per capita, the value of life for Chinese and U.S. citizen is the same. With the value of life for rich people in China weighs more than common people in the U.S.). If the rich 1% of Chinese citizens could spend 6,000–10,000 RMB on insurance pooling, it will be enough for insurance companies in China, let alone for all of the 1.37 billion people in China. This is not a problem which can be blamed on a single individual, some groups or institutions. This is simply a process in development. There are many policy initiatives which has been promulgated for the development of commercial (private) insurance schemes; these policy initiatives do not work very well.

Part II: Inequal distribution of healthcare resources between urban and rural China and smart health solutions

1. **Currently how does the central and local governments share responsibility over pooling for social security funds, pension funds and medical security funds? Why is there different reimbursement rates for basic medical insurance schemes at different locations in China? Is the resource allocation leverage towards more developed regions such top tier cities and east provinces? Is there anyway to improve access for rural residents for healthcare services and pension schemes?**

   For providing medical services and pension related services, the central and local government share responsibilities. For basic medical insurance pooling, the public finance system subsidizes every insured with 450 Yuan. The central government subsidizes poor provinces (Western part of China) in China for about 80% of the 450 RMB per capital subsidy, with the central regions in China receive about 60% for the subsidy; even for affluent areas such as Beijing, Jiangsu, Guangdong and Shanghai, the central government subsidizes about 10% for the 450 RMB subsidy. The central government also pays for most of the pension expenses for urban and rural residents.

	2017
Central Government expenditures for healthcare services	10.76
Local Government expenditures for healthcare services	1434.303
Unit: Billion RMB	

2. **Now with the local government pays for a large share for the healthcare expenditures in China, income inequality lead to higher healthcare expenses in cities and regions near the coast rural regions than middle and western regions. This results in the lower life expectancy other low health related indicators in rural areas. For instance, the life expectancy in rural China in 2015 is about 75.6 years and in urban China about 77.9 years. A lot of rural families have fallen back to poverty when exposed to healthcare costs. In the meantime, patients with critical conditions such as cancer, have to ask for loans from the bank, including asking for donations**

from internet based medical aid platforms. Are there any plans for the National Health Security Administration to improve healthcare insurance coverage for rural residents?

The difference in life expectancy in rural and urban China is not just because of healthcare issues. It is also related to environmental factors and life style choices. The bottom layer of the social security system has covered poor people in rural areas. Currently the social security system in China tries to cover the living, housing, education and healthcare needs for the poor, with the safeguard measures are in place (to reach the target of poverty alleviation, secure relevant needs for food and clothing, and to secure compulsory education, basic medical care and housing). The critical illness medical insurance scheme is trying to lower the payment threshold for critical illness such as cancer and change the reimbursement rate from 50% to 60%.

The nostalgia for barefoot doctors is kind of sentimental. The lack of medical resources for rural residents is due to lack of marketization; marketization cannot be blamed for the lack of medical resources for rural residents. The healthcare industry shall adapt to the development of the economy. Government should only intervene when market cannot resolve the problem. Government policy and other administration tools should play a supplementary role rather than become the main controlling methods.

There are a few doctors willing to work in the countryside. The government offers subsidies for doctors working in rural areas. 80% of the subsidies are paid directly to doctors in rural regions via the public finance department on the county level on a monthly basis. (From literatures, different regions have various subsidy standards. Take the example of Guang Dong province, the government selected the doctor in the countryside and pays about 20,000 RMB per year to doctors serving in the countryside of poor regions.) No one has thought about offering more incentives to doctors so they are willing to serve in the countryside.

Do you think IoT can help to cut medical costs or at least slow down rising healthcare costs?

It requires a lot of capital to develop IoT and its application in the healthcare industry, a type of solution in its early stage of development. A lot of businesses are burning cash for developing new technology. Utilizing big data to perform population management, for instance, to make projections on trends of disease risk is a single case. In the meantime, telemedicine (long distance operation, etc.) can solve high-end problems. AI cannot guide doctors on how they treat patients. The core of the tense relationship between patients and doctors originates from lack of communication. In the current healthcare system in China, the doctors in public hospitals do not have the time communicate with patients online. These work can be done via nurses.

Part III: Value based healthcare system and IoT

1. **What are the supporting policy initiatives for realizing value based healthcare solutions, where the reforms focusing on improving patients experience, making decisions based on data, promoting hospital operation efficiency, lower healthcare cost, and promoting diagnosis prevision rate. Are there incentives for hospital and other institutions to focus more on medical results? Do you think the IoT will help to establish the value based healthcare system?**

   For patients with NCDs (diabetes, hypertension, pregnant women, patients with critical mental disorders), they can sign up for GP service which will continue to monitor their health conditions. Realizing value based healthcare is not a technical issue, with technology only solving the details part of the problem. The administration hopes to realize the value based healthcare services in 5 years.

   Hopefully within 5 years, every family can have a qualified GP to provide services with every citizen has an EHR record. The designated GP to every family and patient will be the core of the services, as then it will be easier to pay by head. With off-site settlement, it is difficult to realize paying by head. The medical equipment manufacturer is in charge of controlling cost with the two invoice system.

2. **The current fee for service system is undergoing some reforms. The reforms propose the settlement based on lump-sum payments based on heads, disease related groups. Is there any chance that the healthcare security fund management administration can control rising healthcare costs, improve service quality, and improve health for the whole population based on the whole healthcare network?**

   DRGs (Diagnosis related groups) settlement is based on evaluating the cost treating the group of the patients with similar patient age, gender, disease diagnosis, treatment methods, and patients' conditions. The healthcare security funds will pay in advance to healthcare institutions to cover the costs of certain DRGs.

   To realize settlement based on DRGs, it is necessary to budget on healthcare security funds. Now the healthcare security administration is sitting at the end of healthcare services, while the administration is planning on becoming the strategic buyer. The plan is to control the total cost and payment to healthcare institutions and make the reimbursement based on historical data. There have been experimental programs running for settlement based on specific patients and DRGs.

In 2018, the State Council promulgated *Notice on application for Pilot Programs of Payment by DRGs*, and began to test the payment model by disease type in some regions and hospitals. In 2017, the State Council issued *Guidance on Further Deepening the Reform of the Settlement Model of Basic Medical Insurance Schemes*. The document pointed out that, by 2020, the healthcare administration aims to use DRGs to pay to all medical institutions and medical services on a national wide basis. To reach the goal, a multi-layer settlement model medical insurance that adapts to different DRGs and characteristics of different healthcare institutions and services should be widely implemented throughout the country.

In the future, the health security administration plans to become the manager to choose medical services supply, and to set medical services price level, to administer the incentive system for healthcare institutions, to supervise medical services quality and to coordinate medical resources allocation. The healthcare administration departments did not administer the medical resources in the past.

Part IV: Aging, Chronic disease management, Population management and smart healthcare

1. **The aging process in intensifying in China. In 2016 the crude birth rate in China is about 12 per thousand (World Bank 2018), with the dependency ration in 2030 projected to reach 25%. What are the policy initiatives targeting at the rising healthcare costs associating with the aging population? How does the National Health Security Administration plan to deal with the projection that by 2024 the social security fund will reach a balance deficit of about 735.3 billion RMB?**

    The medical security fund now has a balance of around 100 million RMB. The medical services and other services are a combination of necessity goods and luxury goods; and is a combination of both rigid and flexible needs. To realize the control of waste for the medical security fund, the core solution would be to reform the settlement methods. The most important part of the medical reforms is to pay for medical services in packages (by DRGs). The reform aims to change the fee-for-service model whereby the security funds pay afterward whereby services have been incurred to one where the fund pays in advance. Other measures also target at medications with uncertain effects and unnecessary check-ups. The administration hope to establish a supply-side oriented mechanism to give incentive to doctors and hospitals to control cost (if the total amount of costs stay the same, hospitals and doctors need to control the costs other than services to optimize their income).

2. **Urbanization has lead to changes to traditional ways of long-term care for elderly. The solution used to be three generations in the same family live together under the same roof; now the smart home care solutions are emerging for the elderly, with smart care solutions focus on self-management solutions for the elderly, aiming to ensure the elderly live independently at home if possible. With sensors and interaction via audio/video, it is possible to leave the elderly to live independently if possible at home. Is there any plans to cover the smart elderly care solutions in the urban and rural employee basic medical insurance schemes?**

   Elderly care is not just responsibility for the family, but also for the healthcare institutions. Some cities are running pilot programs for covering home care costs for the elderly. For instance, Wu Han is drafting 'Home Care Subsidy Solution Guide for the Elderly'. For the elderly who is willing to stay at home, the government is going to pay for the door-to-door services. For the elderly who is willing to stay at care facilities, the government will pay for about 200–800 RMB per month of subsidy depending on the physical condition and income of the elderly. For those with relatively good physical conditions, the government will rely on "Internet + Home Care" model, and to integrate different kind of elderly care resources, and to use government subsidy to offer "assistant for meals, cleaning, medical services, and long-distance care" services. The goals are to support elderly with professional care, and nursing support at home with virtual elderly care facilities.

3. **What are your opinion towards smart city and smart healthcare solutions targeting at population health management, chronic disease management, and communicable disease management?**

   Utilizing smart healthcare to perform health management is a progressive process; ultimately it depends on the effects. Market forces, industrial firms are supposed to lead the development of smart health solutions instead of government. The most important thing is to promote healthy life style for the population.

4. **The amount of patients with cardiovascular disease, hypertension and osteoporosis is growing rapidly in China; this is related to the unscientific diet structure and lack of exercise, and overuse of alcohol and cigarette. Is there any specific policy targeting at the growing population with chronic diseases?**

   There are no specific funds targeting at management of NCD. The basic public health service is targeting at elderly people (for 65 years and older), patients with diabetes and hypertension. The use of alcohol and cigarette is a social problem rather than a medical problem.

CHAPTER 4

COVID-19 and the Digitalization of the Healthcare System

1. Introduction

The previous three chapters discussed the demand for smart health solutions in Europe, China and other developing economies such as South Africa. Subsequently, the new business model emerging in the smart health solution industry were analyzed. The book also used empirical data collected from institutional stakeholders to study the power dynamics during Covid-19 for implementing telehealth solutions. In this chapter, I aim to corroborate the perspective of individual users' towards telehealth solutions, with empirical evidence in Beijing.

The COVID-19 global pandemic has rendered the elderly as the most vulnerable group worldwide. AI has been used to help pharmaceutical companies develop new vaccines as well as predict the subsequent Covid infection trends. For instance, BioNTech recently acquired Instadeep to strengthen its capabilities to deal with Covid 19 variants (Kuchler 2023). Moreover, AI has played an important role in making healthcare less costly and more accessible to the masses. For example, AI has changed how healthcare care data is processed. For patients in China, accessing online pharmacies, hospitals and clinics has become the new norm after the global pandemic. Investments have flown into the healthcare sector (medical devices, pharmaceutical companies, consumer electronics, digital healthcare platforms, etc.), driven by demands from healthcare providers and patients. Saudi Arabia has successfully built an AI and IoT-connected hospital (Masmali and Miah 2021) with technical support from Huawei (Huawei 2020).

ChatGPT has become one of the most exciting developments in the field of artificial intelligence. The emergence of large-scale language and image-based models can be owed to the growing hardware capabilities to handle large amounts of data with GPUs and super computers. Although the cost of processing such data has been lowered significantly, the resource for building such large-scale models is significant. In the next couple of years, it is likely that companies such as Google, Amazon, Microsoft are going to dominate the market. The competition between major tech players is rather fierce; for instance, recently Google finally integrated Deep Mind and the Google AI team to facilitate the development of generative AI and to quelle internal politics. Google, Microsoft have integrated Chatgpt with search engines, as well as with office apps such as Google Docs, Excels, etc.

Moreover, generative AI is much more than large language models; it covers other media outlets such as image generation, video generation, supported by large image and video data bases created by users around the globe. The cost has become lower to train algorithms as semi-supervised models are being used by data scientists to reduce the amount of data in need. Faster and more powerful AI are becoming a reality for our generation, ending the ice age for AI development in the 1980s.

According to Goldman Sachs (2023), generative AI could lift the global GDP growth by 7% (with an economic value of almost 7 trillion) and lift productivity by 1.5% in 10 years. Generative AI have a market size of $150 billion with potential use cases in medial, healthcare and technology (Goldman Sachs 2022).

However, there are indeed risks associated with using generative AI from legal, ethical to regulatory risks. In Italy, for instance, generative AI was banned on a temporary basis or the risk of data leakage. The European Parliament is discussing a legislation and tighter regulation regarding generative AI (Johnston and Espinoza 2023).

With the introduction of generative AI, the life of doctors, researchers, would be much easier so they can focus on the analytical part of the job and leave the more mechanical ones to AI. So far Chatgpt has been integrated into Chrome to help doctors to process patient talk records into prescriptions, consultation summaries, as well as follow up letters.

In this chapter, the main focus will be on the use of digital healthcare solutions for chronic disease management, aging patients care, healthy living, as well as accessing healthcare services. The chapter is structured as follows. Paragraph 2 provides the literature review on research methodologies to study user willingness for telehealth solutions. Paragraph 3 presents the research design for the analysis, while paragraph 4 presents a qualitative analysis of focus group studies and reveals the reasons as to why users choose telehealth solutions over traditional health solutions.

The final paragraph summarizes the main findings and implications of the paper.

2. Use of AI in Combating Covid and other Chronic Diseases Such as Breast Cancer

The Covid-19 pandemic has driven an exponential growth in the use of online hospital and pharma services. One such platform experienced a 198% growth in merely one year during the zero-Covid era.

According to an analysis by Deloitte (2021), there are two types of internet-based hospitals—"Hospital+Internet" and "Internet+Hospital". The "Hospital + Internet" model is often developed by public hospitals; this model has advantages in terms of insurance coverage, policy supervision framework and medical resources, and is on the same level of patient experience as internet-based hospitals. The "Internet + Hospital" model has advantages in information access, platform running, as well as healthcare continuum (diagnosis, treatment, recovery and healthy living) coverage.

Mortality rate analysis shows that the age group of more than 50 years old has a higher death rate than other age groups (Koonin et al. 2020). The reason lies in the fact that COVID-19 has more severe effects on a population with multiple chronic diseases such as hypertension, diabetes and cardiovascular diseases, rather than the healthy sub-population. Health management of the aging population is thus intertwined with pandemic prevention and control.

Given the increased pace of aging and urbanization in the Chinese society, and the lack of high-quality medical resources and trained clinicians, there is an urgent need to look for alternative solutions such as telehealth solutions. The implementation of telehealth solutions faces challenges among elderly users because of their lack of experience with technology and, in turn, lack of trust. Other factors such as household income, education and the health status of the user may also play a role in this matter.

Digital healthcare solutions are transformative for the healthcare system stuck in the paper and pencil stage. The decentralized healthcare model is proven to be less costly, letting patients stay at home and doctors work remotely (World Economic Forum 2023). Homecare expenditures is set to outgrow healthcare expenditures from all other segments, leading to investments flowing into the profitable digital healthcare sector. However, as the digital healthcare model does not require patients to meet doctors face-face, it makes it more difficult for patients and doctors to build connections and trust one another. Moreover, as some of the conditions such as mental health issues require doctors to observe patients up close,

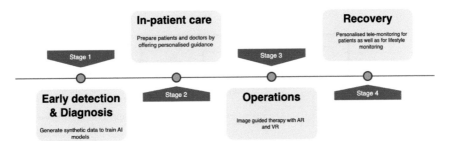

Figure 4.1: The use of generative AI at different stages of care for patients with breast cancer.
Source: Author's illustration.

it becomes easier for patients to lie to doctors over a 30-minute session online, in an attempt to get a prescription. For instance, in the Cerebral case—the lawsuits launched by families against the platform for the over-prescription of the opioid drugs, as well as the subsequent death of teenager users. There were some physicians choose to leave the platform as they feel that the platform is pushing them towards prescriptions without confirming if the patients have ADHD over a 30-minute session (Safdar and Fuller 2022). In fact, to promote telehealth solution platforms such as Cerebral, the marketing campaigns by Cerebral and Done Global Inc. focus on promising ADHD prescriptions to patients (DEA 2022).

The use of generative AI can boost productivity for doctors and help patients to improve their experience at different stages of care for diagnosis, treatment (in patient care, and operations) and recovery. This is illustrated in Figure 4.1.

2.1 Diagnosis

As radiologists in a hospital deal with patients with different types of cancer and sample examinations take time, usually patients are told onsite whether doctors suspect there are questionable results from MRI images and X Rays in China. In a dark room, questionable tissues from mammogram are marked on a screen, generating anxiety for patients. The use of AI in detecting breast cancer can help doctors to reduce the amount of time in need to process the images as well as improve the accuracy for diagnosis, allowing early detection of symptoms for breast cancer.

The use of generative AI can allow data scientists to train the algorithms at a faster rate with less amount of data. There are four types of models typically used for generative AI, namely large language models (LLMs), the generative adversarial networks (GANs), the transformer-based models and the variational autoencoder models (VAEs) (Brady 2023). Figure 4.2 depicts how GAN models work. The model consists of two components, a generator and a discriminator. Both real data and generated data are used to train the model; This allows the model to deal

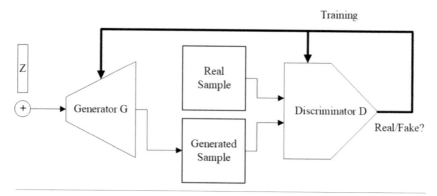

A diagram illustrating how a generative adversarial network works. Image [CC BY–SA 4.0](https://creativecommons.org/licenses/by-sa/4.0/deed.en) האדם-מהשוש on wikipedia

Figure 4.2: Diagram depicting a generative adversarial network. Source: Brady 2023.

with repetitive tasks much better. The adversarial process continues until the data generated by the generator cannot be distinguished from the real data.

VAE models can take a large amount of data and compress it into a small bundle. This can generate new data similar to the original data. Thus, VAEs are often used to generate pictures, audio and videos. In the case of breast cancer detection algorithms, VAEs can be used to generate de-identified medical images (X-ray, MRI images) to help data scientists to train the algorithms, thereby reducing the amount of data required and the time in need for algorithms to reach 85%–90% accuracy (typically required for medical use of algorithms).

Moreover, early detection and prediction of breast cancer can happen with personalized medicine, thus with genetic analysis, as well as with combination of family history and personal lifestyles. This process requires analysis of a large amount of synthetic data, which is not often available due to the limitations created by lack of data interoperability as well as stakeholder collaborations. VAE models can be used to generate such data sets to help to train the model and get the model ready for real-time diagnosis.

Because of sensitivity of healthcare data under GDPR protection— "data minimization" principle, demanding that data controllers to limit the collection of data necessary for a clearly defined purpose (European Union 2018). It is difficult for data scientists to access medical images as the data cannot leave hospitals. Data interoperability is another issue troubling data scientists. As X ray machines and MRI machines used in different hospitals are often generated by different manufacturers (Philips, Simens, GE, etc.) and stored in various electronic health records

(EHR) systems, the difficulty of connecting such systems adds to a different layer of problem for accessing high-quality medical data. The lack of uniformed and clean medical data results in difficulty in creating high-quality algorithms. The use of generative AI can help data scientists to generate new medical images without accessing a large amount of real-patient data.

2.2 Treatment

For breast cancer patients at later stages treatment plans often entail chemotherapy, as well as surgeries. The in-hospital experience can be confusing for patients as well as stressful for both patients and doctors. Chatbots powered by LLMs can help patients to be informed about the necessary information in need for inpatient care, as well as prepare them to change their lifestyles to prepare for the operation. Moreover, for doctors, chatbots can help to summarize electronic health records, and medical images charts from different sources (family doctors, previous hospitals) and fully prepare doctors for the patients when they are admitted. Moreover, generative AI tools can help doctors to prepare for operations as well as for post-surgery paperwork, generating data in a more standardized format to reduce the number of administrative tasks for doctors. This will help them to focus on the training for surgeries and avoid distractions.

2.3 Recovery at home

Furthermore, in later stages of patient care, LLMs can be used to help patients to recover and answer questions. Studies have shown that telemonitoring can help patients to stay connected with doctors, especially when patients are diagnosed with chronic diseases such as cancer, diabetes, cardiovascular diseases, psychiatric conditions (Helissey et al. 2023). Telemonitoring is shown to help to improve clinical outcomes by recording patients' daily activities more accurately, and provide emergency care at falls, and other types of emergency situations (Malasinghe et al. 2019). For patients with breast cancer, dealing with societal pressure from changes in body forms as well as working in supporting groups help them to deal with the psychological challenges arising from chemotherapies as well as post-operative recoveries. Chatbots, powered by LLMs can offer them a channel to divest their emotions, as well as offer helpful counseling advice at challenging situations.

To assess the user preference (or aversion) towards the use of telehealth solutions in a non-clinical setting in China, it was decided upon to use focus group studies for individual users in rural and urban Beijing.

3. Research Methodology

3.1 Ethics approval

Ethics approval was obtained in May 2019 from the committee of University of Macerata. Based on the ethics approval, analytical results from focus group studies have been designed to analyse the stakeholders' attitudes towards whether IoHT solutions can help bridge the gap in the current healthcare system demands in China.

A summary of the data collected and the study design can be found below in Figure 4.3.

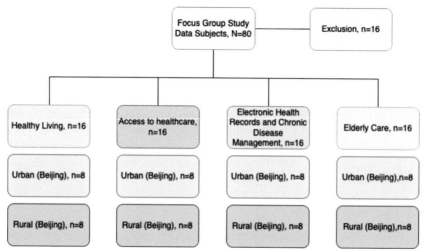

Figure 4.3. Focus Group Study Design. Source: Focus Group Data Collection, Author's illustration.

3.2 Study design

The study was conducted onsite, with the support of University of Chinese Academy of Science as well as Beijing Cinso Consulting Ltd. The study was also recorded, with both video and audio recordings available. In total, 8 focus group studies were conducted over a period of two weeks from the end of March 2019 to the beginning of April 2019. Each group contained 8 participants selected over a candidate pool. Overall, 64 participants were involved in the study, with one coordinator guiding their conversations. One typewriter was used to transcribe the conversation onsite in real time. The coordinator was guided in selecting candidates; the questions for each group were designed by the author and the coordinator was also guided in organizing conversations. Meanwhile, the author sat with the team onsite to monitor the conversation from a one-sided window. The average conversation for each group lasted for 60–90 minutes, generating a transcript of 20–25 pages in Chinese. The participant was compensated for their time with 400 RMB/hour. The entire study cost around 120,000 RMB, equivalent to 17687.37 dollars (2210.1 dollar/group).

Meanwhile, the study proposal was already approved beforehand. The study took a semi-open interview structure. Rural residents are defined as residents of rural neighborhoods (county, township, village level in Beijing), whereas urban residents are defined as residents of urban neighborhoods (city, district level in Beijing).

It is worth noticing that all level III hospitals in Beijing are in the city, with only level II and level I hospitals in rural areas.

The selection criteria for male participants are as follows: in the age group from 20–60 years old, male participants should have an independent source of income and be independent consumers. The selection criteria for female participants are as follows: in the age group from 20–55 years old, female participants should have an independent source of income and be independent consumers. The reason behind a lower age threshold for women is that the legal retirement age for women remains at 55 years old, despite the rapidly aging population.

All the participants must have previously used Internet of Healthcare Things (IoHT) devices such as smart watches or other wearables. The participants in the chronic disease management and electronic healthcare records group were required to have family members (or themselves) suffering from a chronic disease. Each focus group was mandated to have a balanced gender ratio, composed of both male and female members, as well as participants from each age group—20–30, 30–40, 40–49 and 50–60 years.

Group 1 (Urban residents) and Group 5 (Rural residents) discussed 13 questions under the guidance of the coordinator. The questions were related to the use of IoHT solutions for healthy living.

Group 2 (Urban residents) and Group 6 (Rural residents) discussed 11 questions under the guidance of the coordinator. The questions were related to the use of IoHT solutions for access to healthcare services.

Group 3 (Urban residents) and Group 7 (Rural residents) discussed 13 questions under the guidance of the coordinator. The questions were related to the use of IoHT solutions for the aging population.

Group 4 (Urban residents) and Group 8 (Rural residents) discussed 11 questions under the guidance of the coordinator. The questions were related to the use of IoHT solutions for chronic disease management and continuous electronic healthcare records.

The 64 candidates were selected from a candidate pool of 80.

3.3 Research methodology

SPSS Modeler and Python 2.7 were used to classify and parse the focus group study transcripts. Microsoft Excel was used to display and analyze the transcripts as well.

4. Qualitative and Quantitative Analysis of the Focus Group Study Results

4.1 The use of telehealth solution for healthy living

In **Group 6 (Rural residenets),** the qualitative analysis suggested that smart watches, bands and apps remained the most commonly used telehealth monitoring tools among rural residents, while scales and connected air quality control appliances were the secondary preference for rural residents.

Respondents from **Group 2 (Urban residents)** mentioned a few sources for accessing healthcare data—Wechat, Keep, Yue Pao Quan (a social app for jogging enthusiasts), Ai Yun Dong (a training app to stay healthy), Huawei Health, Alipay—Ant Forest (where users can trade steps taken to water digital plants), Ping An Healthcare, Ping An Doctor, Le Dong Li, etc. For more professional healthcare services such as appointments with doctors and medication reminders, many patients referred to public accounts and apps from Level III hospitals in Beijing, such as Jing Yi Tong Hospital, Fu Wai Hospital and Chaoyang Hospital. Many patients found online consultation unreliable. The apps allowed data interoperability within one hospital, for doctors from different departments to follow up on patients' electronic healthcare records. The hardware mentioned included Xiaomi Band, and smart home devices such as air ventilators, smart air condition monitor, smart audio assistant, smart cleaning robots, smart camera, smart toothbrush, VR/AR system, etc. Connected devices were popular among respondents who wished to monitor their health conditions, air quality at home, children, patients at home, etc.

Respondents from **Group 2 (urban residents)** were also willing to pay for personalized healthcare consultation, monitoring apps, as well as online psychological consultation services. The trust level for online consultation was not high. Some respondents found brand names such as Apple and Xiaomi more trustworthy, whereas others found it equally impossible to track data flows after tech companies collected them. Some respondents reported that apps from level III hospitals can allow doctors from the same hospital to view the prescription history. Other respondents used genetic tests to access their genetic risk prediction and analysis.

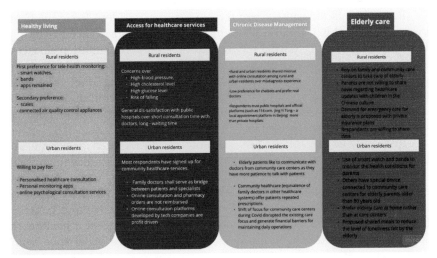

Figure 4.4: Comparison of rural and urban residents over preference for the use of telehealth solutions for healthy living and access for healthcare services. Source: Author's illustration.

4.2 Data sharing regarding the use of telehealth solution for accessing healthcare services

In **Group 5 (rural residents)**, for instance, participants responded with a strong concern regarding high blood pressure, high cholesterol levels and high glucose levels for their parents, along with a risk of them simply tripping and falling. The existence of smart cameras as well as fall warning devices made it possible for the children of the elderly to monitor the health conditions for their parents. The respondents also considered elderly care facilities and elderly homes for taking care of their parents. Regarding the payees for IoHT solutions, the respondents proposed a subscription-based business model for equipment and services.

The analysis of the transcripts of the group also suggested a general dissatisfaction with services provided by public hospitals. This was reflected in the short communication between patients and doctors, high cost of consulting doctors, unnecessary examinations, and crowded hospitals. The dissatisfaction triggered a demand for online doctor consultation services, online doctor registration services and online hospitals. Some respondents mix the concepts of 'family doctor' and 'community doctor' with 'personal doctor'. It is difficult for family doctors to focus on a single patient or one family, whether it is in China, Europe, or the U.S. Moreover, the respondents expected professional consultation from online services and valued the education level and experience of doctors from online consultations.

Respondents are mostly insured with urban resident basic medical insurance (URBMC) or rural resident basic medical insurance (RRBMC) in China, which cover partial medical costs. Online consultation services are not covered under URBMC or RRBMC. Some respondents showed dissatisfaction with online consultation, mentioning the relatively inexperienced doctors. Most respondents wished for continuous health data monitoring and saw online hospitals as the only method of connecting with good doctors and hospitals.

Respondents use telehealth solution platforms such as Chun Yu Doctor, Jing Dong Doctor, Ali Doctor, Jing Yi Tong, 114 for accessing online consultation, Jing Dong Kuai Yao for online pharmacy. These platforms not only make healthcare services more accessible whereby doctors and pharmacists are available 24/7, but also less expensive and benefits both patients and doctors. Patients now have better access for healthcare services and products whereas doctors can gain more experience from remote consulting with patients. However, some respondents reflected on negative experiences from online consulting such as profit driven healthcare services with little care for patient interests. For instance, one respondent who has used Chun Yu Doctor has to join a Wechat group in which doctor will answer consultation for the medical images uploaded by patients with a fee. Another respondent responded by pointing out that doctors and hospitals recommended by Chun Yu Doctor with a fee are not reliable. Other respondents would only use online consultation for minor conditions when they have no time to go to hospital onsite. Online consultation via Ali Doctor and Chun Yu Doctor leads to private hospitals which patients do not trust. Often consultation fee starts small and gradually increase, for instance online consultation by telephone often charge patients by minute.

Some respondents were not willing to share information with their families, but were willing to share information with tech companies, hospitals, and the government, on a de-facto basis. Many patients found it confusing to choose specialists at hospitals, with online hospitals such as Ali Doctor and other platforms; AI was expected to assume the role of the primary care gate keeper, leading patients to the right doctor or department.

For families with members suffering from chronic diseases, monitoring methods were discussed. Camera installation was seen as too invasive by them, whereas fall detection for elderly care was considered as one of the desirable features for remote monitoring.

Respondents from **Group 1 (Urban residents)** expressed dissatisfaction with the amount of medication patients could get per visit with doctors. The other aspect of dissatisfaction arose from the number of medical examinations one patient had to go through for simple symptoms such as cold. With the bed occupancy rate at hospitals remaining high, the bed occupancy time for patients shortened. Respondents complained about leaving the hospital 1–3 days after surgery—when professional care was needed the most.

Some respondents also shared their experiences with community care centers and family doctors. Most patients signed up for the family doctor service at community care centers. The respondents reflected on the higher refund ratio at primary care facilities like community care centers. They believed that the family doctor should work in teams and serve as the bridge connecting patients and specialists at higher level hospitals. One respondent believed that a family doctor should be a retired or experienced doctor. The point raised herein was that oftentimes, inexperienced doctors offer online consultation, through only pictures and texts. Furthermore, online consultation is not reimbursed through basic medical insurance for rural/urban residents; neither are reimbursements made for online pharmacy products.

Regarding the functional design, the respondents suggested that medication reminders can be integrated with online hospital platforms and wearables. Some of these respondents had family members suffering from chronic diseases.

4.3 The use of telehealth solutions for aging population

In **Group 7 (rural residents)**, for respondents relied on family members and community healthcare to take care of parents suffering from chronic diseases, most in the format of regular checkups. Respondents alluded to using home monitoring devices such as blood pressure monitors, cameras, and wearables for elderly care. However, elderly users were not willing to use these devices due to their lack of technical skills; generally, the respondents dealt with these challenges by going to their parents for conducting regular checkups.

The respondents also shared experiences of misdiagnoses by online consultation with doctors from privately owned telehealth solution platforms, and the resulted mistrust in online consultation. When asked about the reasons for the mistrust, respondents suggested the fear of talking with chatbots rather than human doctors. Moreover, these respondents showed a tendency to trust official sources from level III hospitals, as well as from Jing Yi Tong—a state-owned platform connecting level III hospitals in Beijing.

Due to the nature of healthcare services, online platforms which serve as primary care gatekeepers are often under the control of local governments. For instance, Jing Yi Tong is a platform co-owned by several state-owned companies in Beijing. The platform was developed by a startup and funded by a local bank. It enabled over 1 billion patients to register with doctors at hospitals in Beijing. However, the platform suffered financial loss over the years and relied on financial subsidies to survive. It became difficult to sustain after several years of operation.

Respondents from Group 7 discussed commercial insurance membership which includes services such as free annual health checkups. The demand for emergency care devices or services provided to the elderly was proposed by the respondents in this group. The acceptable business model proposed was the freemium business model, with the price of equipment and services ranging from RMB 5000–10000 (USD 736–1472.93). With data sharing, many parents in China are not willing to share bad news with their children; thus, respondents felt that it is the most convenient to share data with the government or tech companies which they trust for a primary care emergency service. Some respondents were willing to share data on an anonymous basis only.

Respondents from **Group 3 (Urban residents)** reflected on their experience of using telehealth solutions for elderly care. One of these respondents mentioned a phone-connected monitor panel at home for elderly patients over 80 years of age. However, the system was only available for the elderly residing in a particular residence neighborhood in Haidian, Beijing. The rest of the respondents reported problems of parents forgetting to take their medicines, reminders needed for parents to monitor blood pressure fluctuations, etc. Some of them had installed smart cameras for parents, while others resorted to smart watches and bands with basic functions such as sleep monitoring, steps tracking, etc.

Some respondents gave positive feedbacks towards the family doctor service at community care centers. The family doctors were reported to share phone numbers with families having elderly patients, in the neighborhood. Elderly patients with chronic diseases received regular calls regarding their conditions, from doctors. It was also mentioned that the elderly patients could receive in-house visits from family doctors, making it easier for those with conditions like hypertension and diabetes to access healthcare services at the primary care level. Respondents who had consulted with online doctors had lower level of trust for them, compared to the community care doctors. The respondents preferred video consultation with doctors rather than relying only on chatting. One respondent recounted their experience with chatbots on the consultation apps, resulting in low trust in such a consultation app. Other respondents also reflected on lack of belief and trust in online consultation services.

The respondents were not happy with the bureaucratic procedures at public hospitals. These procedures made it impossible for patients, especially the elderly, to receive care on their own. About 50% of the patients are covered with basic medical insurance coverage, whereas the rest of the patients are covered with both basic medical insurance as well as private medical insurance, the latter often providing extra care coverage.

Personalized care experience and continuous healthcare recording keeping matters to patients. For patients with chronic diseases, it is important to maintain electronic healthcare records with prescription history and past medical images for doctors to reference. Patients have mentioned that the electronic healthcare records (EHR) are mainly kept with the doctor and patients do not have access to the EHR systems for altering or adding to the records.

The elderly care home and elderly care center model were not welcome by the respondents. They preferred a home-based care model, suggesting that, based on their experiences, elderly care was provided in the community in the form of shared meals. The elderly prefer care in such a format, rather than living elsewhere out of a home setting.

4.4 The use of telehealth solutions for chronic disease management

Respondents from **Group 8 (Rural residents)** indicated that community care centers do not have 24-7 emergency services, driving them to level III hospitals. Most respondents referred to online consultation only for minor conditions. Respondents also expressed interest in services such as home testing kits which would make life more convenient for patients.

These respondents had a low level of trust in sleep monitoring functions and wearables. They reflected more positively on fall prevention services at home and calorie monitoring apps. They also suggested a price tag of RMB 200–1000 (equivalence of USD 28.87–144.36) for fall-prevention tags as a device + service package.

Respondents from **Group 4 (Urban residents)** provided interesting feedbacks towards why community healthcare centers may be welcome by elderly residents. The elderly patients often find doctors from Level III hospitals difficult to approach and too busy to communicate with. On the contrary, family doctors from community care centers can chat with the elderly. Some respondents, however, found it difficult to trust doctors from community care centers, finding them too inexperienced and unable to deal with complex conditions such as pneumonia. The conditions whereby community care centers served patients with chronic diseases required an in-person visit during which the doctors at the community built a healthcare record. Most respondents went to community care centers to get repeated prescriptions rather than for diagnosis; the diagnosis usually happened at level III hospitals. Some respondents recalled their negative experiences at community healthcare centers—about getting different prescriptions compared to those received at hospitals as certain medications are not available at community level healthcare centers.

Community care centers focus on monitoring vulnerable groups in China such as elderly, pregnancy women, chidren, population with mental illness, population with diabetes, etc. For chronic disease monitoring, such as diabetes monitoring, patients often referred to home monitoring devices. Respondents sought to compare data trends in blood glucose and blood pressure levels (for instance) to control conditions such as hypertension, diabetes, etc. However, the entire community care system was disrupted by Covid as the shift of public healthcare system focus has been transferred to Covid testing, quarantine and control of the population. The national healthcare commission has issued new policies outlining the guidance to mark the entire population by green, yellow and red for low, middle and high-risk groups (National Healthcare Commission 2022). This rendered the daily operations of community care centers as secondary, resulting in patients rushing to hospitals whereas community care centers met financial barriers for continuous operations.

Tech companies have created new business models such as offering free insurance coverage to exchange for users' biometrics data and exercise data. The freemium insurance model was mentioned by one respondent; this respondent had signed up for the 'steps exchange with free insurance' model wherein 10,000 steps per day could be exchanged for free insurance. Some respondents reflected on the need for nutrition monitoring for weight control and healthy living. Obesity was reported as one of the main concerns by the respondents in Group 4. However, obesity is not listed as a health condition that community care centers monitor, although it plays a role in other chronic conditions such as hypertension and diabetes.

5. Summary

The comparison between rural and urban respondents did not show significant differences in terms of accessing online healthcare services such as consultation and pharmacy. It seems that respondents from urban areas have better access to elderly care and community care services; respondents from rural areas seem to have used the service more often though. Respondents shared their vision for community healthcare centers and continuous healthcare services as well.

Moreover, the respondents were willing to pay for personalized healthcare services, irrespective of whether they were covered by social insurance (basic insurance schemes) or private insurance schemes. Meanwhile, the level of trust in online healthcare consultation and other online hospital services was low, unless such services were provided by level III hospitals. Considering the fact that these focus groups were studied in April 2019, before the Covid-19 crisis, respondents may have become more accustomed to online consultation during and after the lockdowns.

Building trust with patients, despite their age, career and income is a key for the success of telehealth solution providers. This study has its limitations as the respondents resided in Beijing and had access to abundant healthcare services and relatively high level of community healthcare services. This cannot be said for patients in other cities and regions of China, where healthcare service quality as well as access to healthcare services differ.

The use generative AI has clear benefits in improving the health outcomes for patients. Health economic analysis suggests that compared with no screening, national level screening for breast cancer for instance in the United States can lowers breast cancer mortality and improves quality-adjusted life year (QALYs). Even though the cost of implementing such programs remains high at this stage ranges from $50,223 to $51,754/ QALY (Rim et al. 2019). With the implementation of generative AI at a large scale, the cost for national screening can be lowered as generative AI is good at standardized tasks and lowers the cost for personnel costs.

The implementation of the generative AI in different clinical procedures and scenarios can be met with barriers over the difficulty of integration with existing service providers such as EHR systems. Since medicine has always been an evidence-based profession, the value of generative AI can only be proven when patients and doctors experience such benefits. Moreover, further implementation of generative AI can only be possible with close collaboration of different stakeholders such as patients, governments, tech companies, hospitals, primary level of healthcare service providers.

Acknowledgement

The study received funding from the European Union's Horizon 2020 research and innovation program under the Marie Skłodowska-Curie–ITN Industrial Doctorate (grant 766139) and the National Key R&D Program of China (grant 2017YFE0112000). This book is a reflection of only the author's views, and the European Research Executive Agency (REA) is not responsible for any use that may be made of the information it contains.

References

Almana Hospital: Upgrading medical towers to build a fully-connected hospital. Huawei Enterprise. (n.d.). Retrieved February 22, 2023, from https://e.huawei.com/en/case-studies/enterprise-networking/2022/almana-hospital.

Bishen, S., and Jacquetw, P. (2023). Global Health and Healthcare Strategic Outlook: Shaping the Future of Health and Healthcare. World Economic Forum.

Brady, D. (2023, April 7). What developers need to know about generative AI. The GitHub Blog. https://github.blog/2023-04-07-what-developers-need-to-know-about-generative-ai/

DEA serves order to show cause on Truepill pharmacy for its involvement in the unlawful dispensing of prescription stimulants. DEA. (n.d.). Retrieved February 22, 2023, from https://www.dea.gov/press-releases/2022/12/15/dea-serves-order-show-cause-truepill-pharmacy-its-involvement-unlawful.

Deloitte. (2021). Chinese Internet based hospitals: digital healthcare marches into new age l中国互联网医院：数字医疗迈向新阶段. Beijing.

Goldman Sachs. (2022, November 1). Stability & AI: CEO says AI will prove more disruptive than the pandemic. Retrieved April 20, 2023, from https://www.goldmansachs.com/insights/pages/stability-ai-ceo-says-ai-will-prove-more-disruptive-than-the-pandemic. html?chl=em&plt=briefings&cid=407&plc=body.

Goldman Sachs. (2023, April 5). Generative AI could raise global GDP by 7%. Retrieved April 20, 2023, from https://www.goldmansachs.com/insights/pages/generative-ai-could-raise-global-gdp-by-7-percent.html?chl=em&plt=briefings&cid=407&plc=body.

Gandolf, S. (2023, April 10). Healthcare Marketing and Medical Advertising for Doctors, Hospitals, Healthcare Networks & Pharmaceuticals. Healthcare Success. https://healthcaresuccess.com/blog/case-studies-best-practices/swot.html.

Huawei Global. HUAWEI Global. (2020). Retrieved February 22, 2023, from https://consumer. huawei.com/en/.

Johnston, I. and Espinoza, J. (2023). European Parliament prepares tough measures over use of AI, Subscribe to read l Financial Times. Financial Times. Available at: https://www.ft.com/content/addb5a77-9ad0-4fea-8ffb-8e2ae250a95a (Accessed: April 25, 2023).

Koonin, L. M., Hoots, B., Tsang, C. A., Leroy, Z., Farris, K., Jolly, T., Antall, P., McCabe, B., Zelis, C. B. R., Tong, I., and Harris, A. M. (2020). Trends in the use of Telehealth During the Emergence of the COVID-19 Pandemic—United States, January–March 2020. MMWR. Morbidity and Mortality Weekly Report, 69(43): 1595–1599. https://doi. org/10.15585/mmwr.mm6943a3.

Kuchler, H. (2023, January 10). Biontech buys UK AI start-up InstaDeep in £562mn deal. Subscribe to read l Financial Times. Retrieved February 21, 2023, from https://www. ft.com/content/6670acad-8a5b-4c4a-b6a8-48dc307b6d4d.

Malasinghe, L. P., Ramzan, N. and Dahal, K. (2017). Remote patient monitoring: A comprehensive study. Journal of Ambient Intelligence and Humanized Computing, 10(1): 57–76. doi:10.1007/s12652-017-0598-x.

Masmali, F., and Miah, S. J. (2021). Internet of things adoption for Saudi Healthcare Services. AIS Electronic Library (AISeL). Retrieved February 21, 2023, from https://aisel.aisnet. org/pajais/vol13/iss3/6/.

National Health Commission of the People's Republic of China. (2022, December). Guidance on healthcare work for Covid sensitive population, National Health Commission. Retrieved from http://www.nhc.gov.cn/xcs/zhengcwj/202212/d651c8a88c824cf8911579f503214a44.shtml.

Rim, S. H., Allaire, B. T., Ekwueme, D. U., Miller, J. W., Subramanian, S., Hall, I. J., and Hoerger, T. J. (2019). Cost-effectiveness of breast cancer screening in the National Breast and Cervical Cancer Early Detection Program. Cancer Causes Control. 2019 Aug, 30(8): 819–826. doi: 10.1007/s10552-019-01178-y. Epub 2019 May 16. PMID: 31098856; PMCID: PMC6613985.

Safdar and Fuller. (2022, December 28). Misleading ads fueled rapid growth of online mental health companies. The Wall Street Journal. Retrieved February 22, 2023, from https:// www.wsj.com/articles/telehealth-cerebral-done-ads-mental-health-adhd-11672161087.

Why global health equity begins at home with telemedicine. World Economic Forum. (2023). Retrieved February 22, 2023, from https://www.weforum.org/agenda/2023/01/telemedicine-global-health-equity-at-home-davos23/.

Index